snapper

Published by AUT Media
P. O. Box 7125 Wellesley St
Auckland
www.aut.ac.nz

First published 2008

Designed by Adrian Clapperton at HB Media
Edited by Bradstock & Associates, Christchurch
Printed in China by Everbest Printing

HB Media
PUBLISHING AND PRINT SOLUTIONS

ISBN 978-0-9582829-6-3

CONTENTS

snapper

ACKNOWLEDGEMENTS

A great many people have contributed, in one way and another, to this book. Darren Parsons of NIWA contributed most of the chapter on biology and management, and was very generous in sharing his wide knowledge and research on snapper, contributing to other chapters as well as identifying photographic sources. Ex-fisheries-scientist-turned-publisher Mike Bradstock wrote much of the commercial fishing chapter, as well as the section on the South Island west coast, contributing to several other chapters, supplying illustrations and editing the manuscript with a sure hand.

The design and layout is attributed to HB Media's award-winning art director Adrian Clapperton. Surfcaster Bruce Basher shared his experience and knowledge of Northland's Ninety Mile Beach in Chapter 6. Award-winning Australian journalist and fisherman Steve 'Scoop' Cooper contributed much of the section on snapper on the other side of the 'Ditch'. Jim O'Brien, one of nature's gentlemen, shared his encyclopaedic knowledge of Wellington fishing in Chapter 11, as well as profiling snapper legend Leon Jury in Chapter 9. Prolific fishing writer Tony Orman contributed his knowledge of the Nelson/Marlborough region. Top spearfisherman Darren Shields wrote on his favourite means of harvesting snapper in Chapter 8, and Craig Worthington, one of this country's most observant and innovative anglers, covered the important northeast coast. In most cases I have supplemented these authoritative contributions with my own observations and researches. Other chapters are from my own pen. All of the above have also been generous and diligent in supplying photographs to help illustrate their contributions.

A great many other people and organisations have contributed by supplying many of the illustrations and data used in this work or suffering themselves to be photographed or interviewed. My apologies if I have overlooked anyone.

Alexander Turnbull Library; Archives New Zealand/National Publicity Studios; Auckland City Libraries (NZ), Special Collections; Auckland War Memorial Museum (Catherine Jehly, Andrew Wright); Paul Roux De Buisson; Paul Barnes; Tony and Jean Brljevich; Canvass.co.nz; Carrington Resort, Karikari (Michael and Luciana Nijdam and chef Scott Fraser); Commercial Fisheries Services Ltd (FishServe); Julie-Anne Cunningham; Grant Dixon; Snooks Fuller; Miles Gordon; Bob Guy; Les Hall; Kirk Hargreaves; Hawke's Bay Museum and Art Gallery (Gail Pope); Mark Hemmingway; John Holdsworth and Jeremy McKenzie; Dave Harrison; Dane Hawker; International Game Fish Association; Matt Johnson, Australia; Julien Lajournade, Voyages De Peche, France; Ian Kenny, Norfolk Island; Mangonui Fish Shop; Crispin Middleton; Museum of Natural History, England; Museum of New Zealand, Te Papa Tongarewa (Joanna Moore); Nelson Provincial Museum (Anne McEwan); NIWA; Northern Advocate; New Zealand Fishing News Magazine; New Zealand Post, (Emma Bradley, Allan Hollows); Doug Olander, Sportfishing Magazine, Florida; Keith Patterson, Electronic Navigation Limited (ENL); Iconixart (Sion Wilson); Larry Paul, NIWA/Ministry of Fisheries; Rick Pollock; John Pothan; Reed/Raupo Publishers; Richard Robinson; Pierro Rocco; Carl Scott; Seafood Industry Council (SEAFIC); Stephen Tapp; Darryl Torckler; Steve Waters. And finally, for getting me interested in fish and fishing as a child and fostering that interest into a life-long obsession, thanks to my father Rex Mossman, who had to leave early.

Sam Mossman, 2008

RICHARD ROBINSON

INTRODUCTION

There are certain icons that are closely associated with what it means to be a New Zealander. Like the kiwi, the kauri and pounamu (greenstone). If New Zealand has an iconic sea fish, there can be little doubt that it is the snapper. The noble marlin are revered, but are really summer tourists. I certainly don't mean to denigrate the kahawai or the nuggety blue cod, but more people want to catch snapper than any other kind of fish. This is partly because the distribution of snapper numbers loosely matches the distribution of people in our country – coastal people, coastal fish, and most of both live in the north.

The snapper has many fine qualities. It is a handsome, strong-fighting fish, can be caught using a wide range of methods, and critically, is a well-flavoured white-fleshed top table fish.

Snapper live in a wide range of habitats, from estuaries to deepwater reefs, and can exploit all kinds of food. They are prolific breeders when environmental conditions suit and have the ability to bounce back from over-exploitation – when given half a chance.

Related species are highly valued internationally and farmed in huge numbers in Asia and the Mediterranean. Porgies or sea breams (our snapper are members of this family and not 'true' snappers) make their appearance right through recorded history. In ancient Greece they were sacred to Aphrodite. There are many fanciful stories about the gilt-headed bream; for example Pliny considered its flesh an antidote to poisonous honey, and the sixteenth-century writer John Baptist Porta wrote that it was 'the most lascivious' of all fish and that this 'insatiable desire of the female makes him easy to catch'.

Archaeological evidence shows the importance of New Zealand snapper to early Maori – in some middens in the Far North, the vast majority of fish remains are snapper. Archaeological evidence also indicates a crisis for snapper when the Little Ice Age of a few hundred years ago dropped water temperatures by several degrees – enough to cause the near-extinction of snapper in the South Island and a reduction in their numbers off the lower North Island.

The second great crisis for snapper was much more recent and caused by the greed and mismanagement of man, when uncontrolled commercial fishing in the 1970s and early 1980s brought snapper stocks to the brink of collapse. This commercial excess ushered in the quota management system which, while it has not allowed snapper to fully recover, has at least seen them draw back from the brink of those dark days.

New Zealand's snapper have even featured in international diplomatic affairs: a treaty was signed between the New Zealand and Japanese governments to ensure the withdrawal of the latter's fishing fleets from New Zealand waters where for many years they were catching snapper for the Japanese domestic market.

Today, snapper have never been in a stronger position in the hearts of New Zealanders. Largely this is due to the ever-increasing popularity of recreational sea fishing, with snapper as the most popular target species. The advent of techniques like fishing with artificial soft-plastic lures (to which snapper are particularly partial) has put a big fun factor into snapper fishing which has further increased (if this is possible) the popularity of the species.

So this book is about a fish, and I have tried to keep the scope as wide as possible, drawing in experts from many fields and geographical regions to supplement my own knowledge and researches. It is impossible to get away from the fishing aspect of snapper of course, and while there is much on this topic included, I also hope that non-anglers will get considerable interest and entertainment from this volume.

For me, this has been a fascinating project. I have previously published six books dealing with fish and fishing and found most of the information and inspiration for them in my own head. But this project required some serious research, an enjoyable and interesting exercise from which I learned much, and which brought me in contact with some very knowledgeable and likeable people from all over this blue planet, renewing old friendships and making new ones.

Sam Mossman, 2008

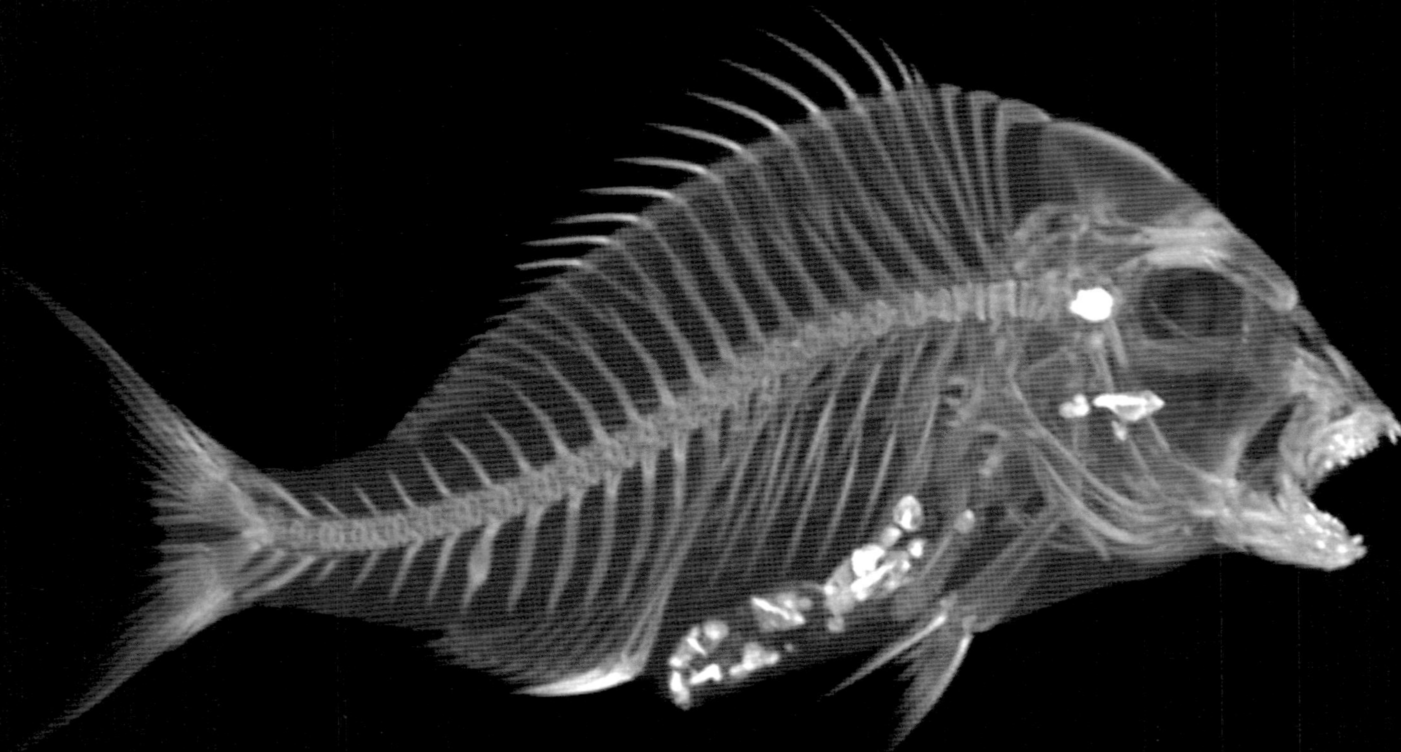

ICONIXART / CANVAS.CO.NZ

SNAPPER BIOLOGY AND FISHERIES MANAGEMENT

01

WHAT'S IN A NAME?

Snapper is undoubtedly New Zealand's best-understood fish species. The earliest record of the study of snapper dates back to Captain Cook's arrival in New Zealand in 1769. Not surprisingly, snapper were caught by the crew of the *Endeavour* soon after arrival in the North Island. While Cook was occupied with observing the transit of the planet Mercury off the eastern coast of the Coromandel Peninsula, the botanist Daniel Solander took a snapper specimen that would be painted by artist Sydney Parkinson and later formally described in scientific fashion.

Much of the initial study of snapper revolved around where to place this species among all the others known to science, and providing it with an appropriate name. While the crew of the Endeavour referred to these fish as 'breams' (because of a fish with similar appearance in Europe), in 1770 Solander gave the fish the scientific name *Sciaena lata*. This would initiate 220 years of taxonomic descriptions during which 20 different scientific names were applied to the fish we know as snapper.

The most recent of these changes occurred in 1990. The numerous scientific names that had been applied to snapper throughout its range – which includes western and eastern Australia, Lord Howe and Norfolk Islands, the Philippines, Indonesia, China, Taiwan and Japan – were discarded and brought under the one name *Pagrus auratus*. The translations of most of the scientific names refer to the golden colour of the fish.

SCHNAPPER OR SNAPPER?

The common name has also undergone some change. In the early 1800s two German scientists, M.E. Bloch and J.G. Schneider, were investigating the taxonomy of many fish species, including snapper. By this stage Europeans in New Zealand had probably coined 'snapper' as the common name, possibly because of the fish's strong jaws or the similarity to the 'true' snappers (which belong to a different family called the Lutjanidae). As well as providing a new scientific name, it seems likely that the two German scientists formed the variant spelling 'schnapper' which is commonly seen in older publications and would reflect the German accent. Other common names include 'bream' or 'brim', which some fishermen from the South Island use to refer to small snapper. Maori also had different names for the fish depending on size, tamure referring to adults and karati meaning small or juvenile fish.

SQUIREFISH

Things became further complicated in 1993 when the American-based International Game Fish Association (IGFA) officially recognised snapper as a gamefish. The IGFA sought to avoid confusion with already-listed members of the 'true' snapper family, which also have powerful jaws and look similar to our snapper, but are not closely related. The IGFA adopted the Australian term 'squire' and listed the fish as 'squirefish', much to the annoyance of New Zealand and Australian anglers. After protests, the fish was finally listed as 'Snapper (squirefish)' in the record books.

This Sydney Parkinson painting of a snapper caught from Captain Cook's ship *Endeavour* in New Zealand in 1769 is the first European illustration of the species. *BY PERMISSION OF THE NATURAL HISTORY MUSEUM OF LONDON, ENGLAND, 9161*

EURO SNAPPER

Snapper belong to a family of fish known as the Sparidae. This is a Greek word meaning 'golden head', a fitting description for our snapper. There are about 110 species of sparids throughout the Atlantic, Pacific and Indian Oceans in temperate and tropical waters.

The European species most similar to our snapper is the gilt-headed bream or giltpoll; in Greek, the name is tsipoura; in Italian, orata; esparido to the Spanish; in Tunisian, jerraf or ourata; in Turkish it is cipura. Most of the names more or less mean 'golden', in reference to the colour of the head, and even the name bream comes from the proto-Germanic, meaning to glitter, sparkle or shine.

The gilt-headed bream repeatedly crops up in the Classics and throughout history. In ancient Greece it was sacred to Aphrodite and there are many fanciful stories about this fish: for example Pliny considered its flesh an antidote to poisonous honey.

Gilt-headed bream were among the earliest of fish to be farmed in the Mediterranean, long before humans discovered how to artificially spawn fish in captivity. Conveniently, the young wild fish enter estuaries, coastal lagoons and ponds, which enabled them to be easily trapped and held until they grew to a good size. It was not until 1981 that Italian fish-farmers first successfully reared gilt-headed bream from eggs; today tens of thousands of tonnes are produced per year, eclipsing the catch of wild fish.

SPAWNING BEHAVIOUR

In New Zealand, snapper spawning occurs between October and February, and the fish spawn a number of times during this period. While direct observations of snapper spawning are rare, observation of fish in captivity suggests that groups of snapper undertake courtship behaviour near the sea floor. A female will then rush to the surface, releasing her eggs, with any number of males following to fertilise them. Snapper are capable of producing prodigious quantities of eggs: a 50 cm fish can release more than 100,000 in one batch, and may do this a number of times over the spawning season. We also know from egg surveys done with fine-meshed nets that some spawning takes place on most days during the spawning season, peaking early in the afternoon.

Snapper start life as small eggs of about 1 mm diameter. The fertilised egg is buoyant and will float near the surface for a couple of days before hatching into a larva. Snapper larvae are rarely seen, but are thought to be bottom-living, and feed on plankton for up to a month before metamorphosing (changing shape) into juveniles or 'baby' fish which look like perfect miniatures of the adults. The growth and survival of both larvae and juvenile fish depends largely on the sea temperature during the first few months after spawning (February to June), with warmer summers resulting in better survival. Water temperature each season therefore plays a very important role in determining how many snapper will be available to catch a few years later.

A common member of the true snapper (Lutjanidae) family is the red bass *Lutjanus bohar*, here photographed on the Great Barrier Reef, Australia. *SAM MOSSMAN*

The onset of spawning is also largely regulated by water temperature. It usually begins at about 15°C, and gradually as the water temperatures rise the fishes' gonads enlarge to produce eggs and sperm. Spawning is also affected in a complex and incompletely understood way by other conditions in the sea. This means that the timing and magnitude of spawning events can vary greatly between years.

Location can also influence spawning events. Traditionally, snapper in Doubtless Bay, the Bay of Islands, Hauraki Gulf, the northwest coast of the North Island and Bay of Plenty may start spawning in November and continue right through until March. Further south, in places like Hawke Bay, South Taranaki Bight and Tasman Bay, the critical water temperature for spawning is reached later and does not last so long, meaning that spawning may only take place from late January to mid-February.

These are generalisations of course, and patterns may differ from year to year with variations in overall ocean warming caused by currents, weather patterns, and so on. One effect is to make the southern snapper spawning season shorter and less reliable. This means that these southern fish are slower to recover from overfishing than those in the Hauraki Gulf, where both commercial and recreation pressure are intense but warmer water conditions are stable and of long duration.

The largest snapper populations, both in New Zealand and in Australia, are nearly always centred on large bays and gulfs which provide core spawning grounds for the surrounding population. There are many theories explaining this, including the possibility that eggs may be retained within these enclosed waters and that the expansive sheltered shallow water may provide suitable nursery habitat.

GENDER BENDERS

Snapper all begin life as females. By the end of their first summer they are around 5 cm long, and over the next two years they grow to about 15 cm. During this time about half of them change sex to become males. Changing sex like this is a quite common among fish (another NZ fish that does this is the spotty or paketi.) While snapper are small they feed on a wide range of small animals including crustaceans and polychaete worms, and live in a range of habitats from estuaries and shallow soft-bottomed coastal areas to rocky reefs. Many of them fall prey to larger fish including john dory, kahawai, kingfish and scorpion fish. As they grow in size snapper become less vulnerable to predators and can tackle a wider range of prey. Eventually they are big enough to prey on fish, but it is not until they are large adults that fish become a major part of their diet.

Snapper reach sexual maturity around about the same time they reach the minimum legal size (25 or 27 cm). Adult snapper occur over all bottom types including sand, mud and rocks, and are found in harbours, estuaries, bays, open coasts and around offshore islands to a depth of about 200 m, though they are much scarcer in the deepest part of this range.

Known range of juvenile snapper

Known range of spawning snapper

Limit of snapper distribution. They are much more scarce in the lighter shaded area

TOP: This sea bream is from Oman waters in the Indian Ocean and features the humped head also found in some Australian snapper. *JULIAN LA JOURNADE, VOYAGE DE PECHE, FRANCE*

ABOVE: An American snapper relative is the sheepshead, a shellfish eater popular with American anglers in the Gulf of Mexico. *DOUG OLANDER, SPORTFISHING MAGAZINE, FLORIDA*

DISTRIBUTION

In New Zealand, snapper are most abundant in the north, with decreasing numbers south of East Cape on the east coast and Taranaki on the west coast. In the South Island the snapper fishery is centred around Tasman Bay and the Marlborough Sounds. In the cold winter months, many snapper move out of the shallower waters of inshore harbours, bays and channels. This is because in shallow water, more of the water mass is exposed to the surface, so it cools down more. Although most of the fish spread out in the deeper, slightly warmer water of the inner continental shelf and down to about 150 m, some remain inshore. At this time many fishermen target larger snapper in shallow areas of heavy foul.

Cold temperatures also affect snapper behaviour (and catch rates) in other ways during the winter months. As snapper are ectothermic ('cold-blooded') they tend to become less active and require less food when it's cold. Consequently anglers report that 'bite times' can be brief – at first and last light, and periods of tidal flow change – with fish being picky and tentative in the way they take a bait. It therefore becomes difficult to draw any real conclusions about the abundance of snapper simply from catch rates.

These changes have recently been assessed with a towed camera system that sneaks up on the snapper 'sleeping' on the seafloor at night, allowing the observer to count individual fish as the camera moves along. Results largely agree with the observations of fishermen: there seem to be fewer snapper inshore during the cooler months and more in spring as the adults move inshore again to form schools on the spawning grounds.

Surprisingly, however, this survey found that the number of juveniles (fish less than 25 cm long) remained constant in these inshore areas during all four seasons. Why fishermen experience plague proportions of small snapper picking their baits off their hooks during summer is therefore unclear. Perhaps these small fish are not very hungry in winter.

WHAT DO THEY EAT?

Snapper eat just about anything, and this is one of the main reasons why they are among the most common coastal fish in the north. They occur across a range of habitat types and are total opportunists when it comes to feeding, so they can exploit many food sources. One study found 99 different kinds of food in the gut contents of a sample of snapper. Nearly half of these were crustaceans such as crabs, with worms, echinoderms (sea urchins and brittle-starfish), shellfish and other fish making up most of the rest of their diet.

The chewed-off arms of brittle-starfish eaten by snapper are often mistaken for worms, and areas where they brittle-starfish are common are often known as 'worm beds'.

While studying the gut samples it has been noticed that snapper caught in the morning and evening tend to have the fullest guts. This corresponds to the dawn and dusk 'bite times' observed by fishermen. One theory to explain this has to do with the changing light conditions. At dawn and dusk a transition of daytime-active and night-time-active animals occurs. As a result the animals are highly active and yet light levels are not ideal for the vision of either group. This may make it easier for the snapper to catch their prey, and would explain why the fishing tends to be best at dawn and dusk.

Another interesting aspect of the snapper diet is the small proportion of other fish – less than 10% – present in the gut contents. This seems strange considering that baits used to catch snapper are often fish of some kind. Despite the small overall percentage of fish in the diet, during spring snapper guts may be full of pelagic fish such as pilchards, as well as other pelagic organisms such as salps (jellyfish-like creatures). This suggests that snapper may feed in midwater at this time of year, and may be related to other things that we know occur at this time such as schooling and spawning.

BELOW: Bream have featured throughout recorded history. This fish plate, held by the State Hermitage Museum in Leningrad, is attributed to a Greek painter from Larnaca (c. 350BC) and features a number of fish species including a bream (lower right).

SNAPPER MOVEMENTS

The movement of snapper is still poorly understood. The best way to investigate fish movement is by tagging them and waiting for fishermen to return the tags. Studies like this have been conducted on snapper many times, with tens of thousands of snapper tagged since the 1940s. However, the movements of snapper remain poorly understood, largely because no studies have shown much in the way of a consistent pattern of movement.

While results vary, there are several factors that are consistent. The majority of snapper movements were limited to a range of about 25 km from the point where the fish was tagged. Some snapper moved much longer distances, with the longest being 418 km. Of the fish tagged in the Hauraki Gulf (where much research has been done), movements occurring in spring seemed to be consistent with travel to areas where the fish aggregate for spawning.

In recent years snapper have been tagged with acoustic tags. Tagged fish are tracked using a hydrophone to listen for an acoustic 'ping' emitted by the tag so they do not need to be recaptured, and this provides a high level of detail on movement. This method was used to demonstrate that some snapper remained within a radius of only a few hundred metres or less for many months, if not permanently.

SCHOOLIES AND KELPIES

As a whole, the results of these tagging studies agree with the ideas of fishermen who believe that school fish move large distances and 'kelpies', 'moochers', or 'old man snapper' tend to reside in one area. The appearance of these two types of snapper differ, with school fish often having a pale or bright pink skin colour and sharp teeth, while kelpies have a darker red or brown skin colour and a dark line on the posterior margin of the tail. Kelpies also often have blunt teeth and sometimes a large bump on the head.

There are many theories about these differences in appearance, but it remains uncertain as to whether they

TOP: These two fish typify the differences in colour and body structure between a 'kelpy' or 'old man snapper' (bottom) used to living in heavy reef structure, and an open-water dwelling 'school' fish.
SAM MOSSMAN

CENTER: Snapper teeth, housed in powerful jaws, are of a generalist design, equally suited to seizing prey or grinding up shellfish. *SAM MOSSMAN*

BOTTOM: Daurade grise, one of many snapper relatives, is a very popular fish along French Atlantic coasts. *ALBAN CHOINIER*

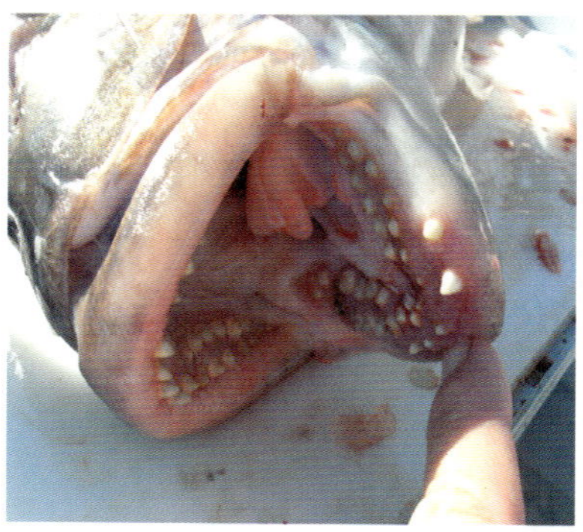

reflect a difference in movement behaviour or lifestyle of the fish. For example, snapper grow more than one set of teeth in their lifetime, so they can replace bluntened teeth even if they have been feeding on hard or abrasive food sources. Snapper skin colour is also affected by several factors and it can change dramatically in a period of weeks.

The appearance of humps on the head is caused by the development of one of the skull bones. Only a few New Zealand snapper develop this hump, but it is relatively common, even in some modest-sized specimens, in parts of Australia.

MARINE RESERVES

In recent years marine reserves have been a focus of research highlighting the impact that fishing has on snapper populations and the important role snapper play in balancing the wider marine ecosystem. There are now many marine reserves in New Zealand, the oldest being the Cape Rodney to Cape Okakari Marine Reserve (usually known as the Goat Island Marine Reserve), near Warkworth, established in 1975.

At Goat Island Bay visitors can swim among groups of large snapper that live in the reserve. There, there are about 14 times as many legal-sized snapper as there are in nearby fished areas and the average fish is about 15 cm longer.

Similar patterns have also been observed at the Tawharanui, Hahei and the Poor Knights Islands Reserves. At the Poor Knights Islands, snapper populations increased by around 300% within a year of the area becoming a fully protected reserve. Many of the snapper observed in the reserve were also large adults, suggesting that previously transient snapper had become residents of the reserve. It is unlikely that this was caused by their being attracted to the reserve, but rather by the accumulation of fish that suddenly were not being caught any more.

The large numbers of snapper in marine reserves also resulted in higher levels of predation on the animals that snapper like to eat and, for example, may have contributed to the considerable reduction in numbers of small reef fish noted in the Poor Knights reserve in recent times.

One noticeable change at Goat Island is that the numbers of sea urchin or kina have been greatly reduced. This may be as a result of predation by abundant snapper and crayfish. Kina are grazers that feed on marine algae such as kelp. As kina numbers declined, areas cleared by kina grazing were replaced with forests of kelp. While other factors such as the presence of a disease specific to the kina and the effect of other predators can modify this pattern, the predatory influence of snapper on the rocky reef ecosystem is clear. Research on processes like this and the detailed level of understanding they provide can only be made by having fully-protected marine reserves to provide baseline information. While the influence of human visitors hand-feeding the fish can modify their behaviour inside these reserves, such activities are mostly restricted to small areas and only occur in some of the reserves. Most studies within marine reserves account for this by including study sites where no fish-feeding occurs.

POSSIBLE EFFECTS OF CLIMATE CHANGE ON SNAPPER

As mentioned, water temperature markedly affects the survival of snapper eggs and larvae. A temperature of 18°C offers reasonable survival rates, while 16°C can mean greatly reduced spawning success. Temperatures above 18°C can mean extra-strong year-classes, and these can be tracked as a bulge in regional population pyramids as they grow.

The average New Zealand sea temperature has risen by 0.7°C since 1800. Regardless of the causes of the present warming trend (a natural cycle coming off the back of the 'little ice age' or the 'greenhouse effect' – take your pick), the effects of increased water temperatures on snapper spawning success are likely to be positive, allowing greater spawning survival and increasing the southern range of the species. Other postulated effects of climate change, such as increasing water acidity and its effect on plankton, algae, shellfish and so on, are unknown. Snapper, however, are a flexible species, able to rapidly adapt to and exploit changing food sources – an attribute that is a big advantage for them over more specialised species.

BELOW: A useful way to investigate fish movement is by tagging fish, releasing them, and getting fishermen to return the tags. This Hauraki Gulf fish was tagged at the Noises on 6 December 2006 and moved a straight-line distance of 20 nm to be recaptured at Maori Rock, near Kawau Island, by visiting Australian angler and rally car driver Michael Guest on 14 May 2007. *SAM MOSSMAN*

LEFT: Kina are a regular food item for snapper. *DANE HAWKER*

BOTTOM: Baited underwater video cameras are one technique used to estimate a snapper population in a given region. *PAUL ROUX DE BUISSON*

SCIENCE AND FISHERIES MANAGEMENT

To assess the health of a fishery and set appropriate catch levels, a lot of information is needed:

• Individual stocks of fish are identified. An individual stock is a group of fish that breed among themselves only, so they can be treated as one unit. The boundaries of a stock can be determined by comparing genetic diversity or growth rates between different areas, and by observing any natural boundaries to fish movement by tagging studies.

• The catch taken from each stock must be estimated. For the commercial fishery this is easy because commercial fishermen are required to report all the fish that they catch.

• Estimating recreational catch is much more difficult. In the past telephone surveys were conducted and some of the interviewees were asked to keep a diary of their fishing activities. Results from this method were unreliable and it is no longer used. Since 2003 the recreational snapper catch has been estimated by a combination of aerial over-flights that count fishing boats, and simultaneous boat-ramp surveys that observe catches being landed.

As a result the recreational catch of snapper is better understood than that of any other species in New Zealand. For example, it is estimated that more than 2000 tonnes of snapper are caught every year by recreational fishers in the SNA1 management area alone – almost half of the commercial take for the same area.

• The rate at which the fishery is replenished by juvenile snapper growing to legal size is also important. This is called recruitment, because once the fish are big enough to catch they become new 'recruits' available to catch. Predicting recruitment before it happens is crucial to the management of the fishery, as it tells us how many new fish there will be. This can be done by conducting a trawl survey each year, where a research vessel tows a net in a standardised fashion (same places, same length of time, etc). The number of juvenile snapper in the catch can then be used to gauge the number of new snapper resulting from earlier spawnings.

This number varies greatly from year to year and is largely determined by the sea temperature over the summer period

BELOW: The life history stages of a snapper
LARRY PAUL, NIWA/MINISTRY OF FISHERIES

1mm egg

egg at 45 hours ready to hatch

1 day old larva 1.7mm

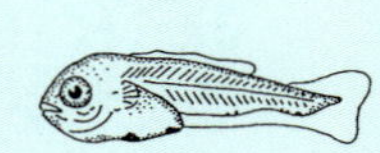

6mm larva

in which the fish were spawned. In years with warm summers, larval snapper that have just been spawned are more likely to survive (partly because the warm water provides plenty of planktonic food for them to eat). Once the number of newly-spawned snapper is determined, the amount of time before they grow to legal size also needs to be estimated. Growth rates of snapper vary between years and between locations. For example, fish from the west coast of the North Island grow to legal size (25 or 27 cm) in 3–4 years while a fish from the east coast will take 5–6 years.

The age of a snapper can be determined by extracting its earbones (otoliths) and counting the rings inside the bone, in exactly the same way a tree is aged by counting its annual rings. Thousands of snapper are aged in this fashion every year to determine growth rates and monitor the age composition of the various stocks. A significant loss of the older age-classes would suggest over-fishing.

• The total weight of snapper in an area (biomass) must be estimated. To get an idea of whether biomass is increasing or decreasing, scientists measure how easily snapper are being caught by commercial fishermen. The more snapper present, logically the easier it will be to catch more of them. The way they measure this is in terms of Catch Per Unit Effort (CPUE). For example, the catch of commercial longliners may be expressed as the number of snapper caught per 100 hooks that were set.

The best method, however, is to conduct a tagging experiment. Thousands of snapper are tagged and released, taking care to ensure they are well mixed throughout the stock by tagging in as many different places as possible. The biomass can be estimated from the ratio of tagged to untagged snapper that are then caught. Once this ratio is known it can be used to scale, from the number of tagged fish that were originally released, an estimate of the number of fish in the stock.

In recent times electronic 'PIT' tags have been used for these experiments. They are about the size of a grain of rice and are implanted inside each snapper. They are retrieved by scientists who visit fish-processing sheds, where an electronic detector is used to identify tagged fish and read off their tag numbers.

Scientists use computer models to predict the growth or decline of the snapper population, based on all these factors. Different scenarios can also be simulated, predicting what would happen to the population depending on different catch levels. This information is then used by fishery managers to decide on appropriate quota limits that ensure the fishery is being fished sustainably. At the time of the last assessment, for example, there were an estimated 65 million adult snapper in the SNA1 area. This is thought to be below the optimum productivity level for the stock, or what fisheries scientists call its Maximum Sustainable Yield (MSY), but the population is expected to rebuild over the next 20 years.

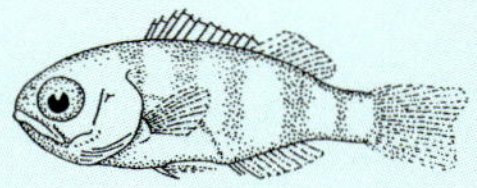

14mm postlarva

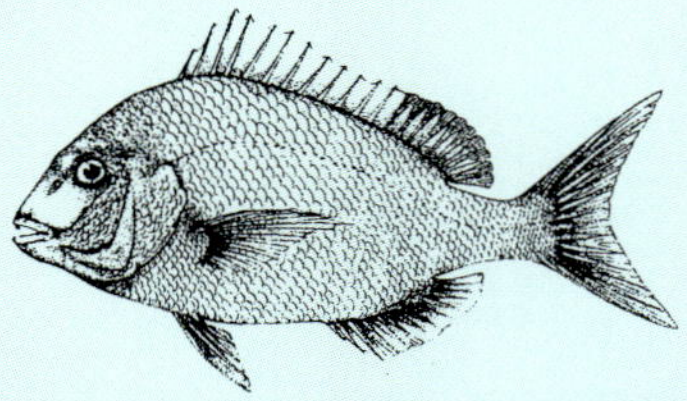

Body shape typical of small to medium sized adults

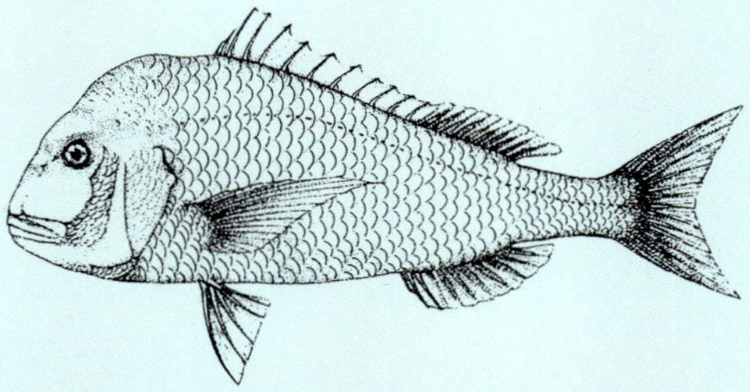

A large adult about 75cm long, with the heavy head and elongated body typical of old snapper

RICHARD ROBINSON

TANGATA WHENUA AND TAMURE

02

MYTHS AND LEGENDS

Fish and fishing have always been important to Maori communities. Their ancestors, the Polynesian seafarers who began to spread out across Oceania around 1200 BC, drew much of their sustenance from the sea and when they reached Aotearoa around 1300 AD they adapted and developed their knowledge and techniques to the new land.

Many Maori myths and legends concern fish and fishing. The Maori god of the sea and ancestor of all fish is Tangaroa whose grandson Ikatere went to sea, where he and his children became fish (ika). One legend involves the legendary navigator Kupe discovering New Zealand while in pursuit of a giant octopus.

The most famous legend, and one known by all New Zealanders, is how the demi-god Maui fished up the North Island.

Maui was the youngest brother of the family, and a trickster who annoyed his elder brothers so much that they refused to take him fishing with them. When they left the shores of Hawaiki, Maui stowed away under the floorboards, showing himself only when the waka (canoe) was too far out to return.

When the brothers refused to give him bait, he struck his own nose and smeared the blood on his magic fish hook, made from the jawbone of his ancestor Murirangawhenua. Casting it into the sea, he caught a great fish, battling it to the surface despite the pleas of his terrified brothers. The fish became the North Island – Te Ika a Maui (the fish of Maui) – with its head to the south. Wellington Harbour is its mouth, Lake Wairarapa its eye, Lake Taupo its heart (to Tuhoe, the Urewera Mountains are the heart), Taranaki and the East Coast are the fins and Northland forms the tail. The hook forms Hawke Bay, with Cape Kidnappers the barb. The waka Mahunui became the South Island – Te Waka a Maui (The Canoe of Maui), and Stewart Island its anchor stone. It is fitting that the creation myth of Aotearoa should involve the biggest catch ever made.

SNAPPER AND ANCIENT MAORI

Fishing provided a rich, dependable and sustainable food source for pre-European Maori and developed into a sophisticated system based on the experiences of generations of fishermen and governed by a tradition of ritual restrictions (tapu) and penalties.

The Maori name for snapper is tamure, a name still used in many other parts of Polynesia for similar-looking but unrelated species of fish. There is no doubt that snapper were of considerable importance to early Maori, particularly to tribes from Tasman Bay to North Cape. Archaeological excavation sites in these areas show heavy concentrations of snapper remains.

The anthropologist Foss Leach has done a huge amount of work in the archaeo-zoological field, and drawn some interesting conclusions in his book *Fishing in Pre-European New Zealand* (2006). Snapper represent on average nearly 16% of fish remains found in archaeological sites, which by this measure ranks them as the third most important fish to

This painting by Charles Heaphy shows a seasonal Maori fishing camp of the 1850s with waka and fish-drying racks, at Rangitoto Island. *ALEXANDER TURNBULL LIBRARY, WELLINGTON, NZ. C-025-002*

ancient Maori economy, surpassed only by barracouta (25%) and blue cod (16.6%). This varies with the area of course, and in the north snapper are easily the most common at 74%, with some sites showing them making up close to 100% of the total fish remains. Snapper bones are numerous in middens around Wellington and Tasman Bay in the north of the South Island, and a few occur in archaeological sites near modern-day Christchurch, Dunedin, the Catlins and even in Fiordland. However, it is in these southern sites (and at the Chatham Islands) that barracouta and blue cod are dominant.

At archaeological sites with extensive snapper remains there are differences in fish size. Near the Hokianga Harbour, for example, the remains show a lot of little snapper, some as small as 10 cm and many under 30 cm. It is suggested that they were caught with a beach seine net, suited to the sandy nature of this area, although to this day the inside of the Hokianga Harbour is known for having large numbers of small snapper, so the habitat may be equally responsible.

By contrast, a site on Mana Island near Wellington showed the remains of many larger fish. This suggests that they were line-caught, which matches the fact that the rocky nature of the area was much less suitable for net-fishing. At both the above sites there does not appear to be much of a change in the size of snapper found in the middens through time, suggesting that there was little pressure on the stocks. However, comparison of archaeological evidence with present-day trawl data from the same regions suggests that the average size of snapper from middens was about 13 cm longer and 2.2 kg heavier than modern-day fish.

The largest snapper remains found in a midden were those of a fish 1.01 m long, which would have weighed about 15.8 kg (35 lb). This is a little smaller than the biggest reliably recorded snapper of modern times, a 1.05 m fish weighing 17.2 kg (37.9 lb) caught at Motiti Island in the Bay of Plenty in 1992.

Leach made an interesting estimation from archaeological and nutritional data. He calculated that each person would have required 2150 calories per day to function. Isotope analysis on human bones from similar areas suggests around 22% of their diet was fish. The minimum amount of snapper remains in a northern midden was 48% (others had as high as 92%). After figuring out the average weight of these snapper, the percentage of the body eaten and the energy content, Leach estimated that Northland Maori ate at least 46 snapper each per year. Given an estimate of 15,000 Maori in the region, this approximates a low-end total of around 2000 tonnes of snapper harvested each year, not counting fish preserved and traded with others. Although there is a wide margin of error involved, these figures certainly indicate the importance of snapper to early northern Maori.

MAORI SNAPPER FISHING TECHNIQUES

With an abundance of fish and a reliance on natural materials such as wood, stone, bone, shell and natural plant fibres, the pre-European Maori developed relatively sophisticated techniques for fishing in waters up to 100 m deep. It is likely that most snapper fishing was done with nets, and hook and line to a lesser extent. The records of early European explorers indicate that by the late 1700s, communal net-fishing was the most common way of catching snapper for the whanau.

Barthelemy Lauvergne captured this scene of local Maori and French sailors hauling a seine net (kaharoa) on the beach at Russell, Bay of Islands, in 1831. *ALEXANDER TURNBULL LIBRARY, WELLINGTON. B-098-005*

NETS (KUPENGA)

An entry in Banks's journal on 4 December 1769 written in the Bay of Islands records a visit to a village there:

'...and after having a little laught at our seine [net], which was a common kings [navy] seine, shew'd us one of theirs which was five fathom [10 m] deep and its length we could only guess, as it was not stretch'd out, but it could not be less, from its bulk, than 4 or 500 fathom [800 or 1000 m]. Fishing seems to be the cheif business of this part of the countrey; about all their towns are abundance of netts laid upon small heaps like hay cocks and thatchd over and almost every house you go into has netts in it making.'

Maori nets (kupenga) used a similar netting knot to that of Europeans, but it was easier to tie. Nets used for snapper were of two main types: gill nets (set nets), specifically sized to entangle target species, and seine nets (kaharoa).

The use of large seine nets in the north and Bay of Plenty suggests most snapper were caught along open sandy beaches, where nets were not at risk of becoming snagged in rocks. These seine nets were often very large – up to a kilometre long and six to ten metres deep. They had small pouches along the bottom that contained stones (karihi), and light wood (often whau or houama) or pumice floats on the top edge of the net. The mesh size varied throughout the nets, with larger mesh on the edges and smaller mesh in the middle to improve net performance.

Most large nets were made from green flax (harakeke) strips. The stiff strips of green flax were partly dried before use to prevent the knots from loosening after the net was made, removing the need for a net needle. Smaller nets, however, were sometimes made of two-ply dressed flax fibre.

Netmaking was a communal activity, carried out under strict tapu, as was the net's first use. The last of the great kaharoa was recorded as having been made at the Bay of Plenty settlement of Maketu in 1885, under the direction of Te Pokiha Taranui, chief of the Ngati Pikiao. Over 300 people were involved in its construction over a nine-month period. It was constructed in seven sections which were then assembled and joined at Otumakaro Flat, below Maketu. The resulting kaharoa was 2,090 yards (1,910 m) in length and estimated to weigh two tonnes when dry. Two large waka were borrowed from the Ngati Rangitihi (another Arawa iwi further down the coast) and joined together to carry the net, with a crew of thirty.

On New Year's Day 1886 the great kaharoa was taken to sea. Tohe te Whanarere oversaw the fishing operation, and after first letting several shoals of fish pass, called out 'Haukotia mai!' (Surround them!) The massive net was set and the waka came back to shore with several hundred metres still uncast, near the mouth of the Maketu River. The weight of the catch was monumental: even a thousand onlookers could not pull the net in. The spare ends of the seine were doubled back to reinforce the centre, and the bottom of the net was raised twice to allow a large part of the catch to escape. Eventually, on the top of the tide, the net was hauled in as far as possible and tied to heavy posts until the tide went out.

LEFT: Maori fishermen and their families drying snapper in the sun before packing into flax kete, Whangaroa, 1902. Large group fishing activities were in decline, but Maori continued to fish to sustain their families. *AUCKLAND WAR MEMORIAL MUSEUM 2159*

BELOW LEFT: Snapper have always been an important food source for Northland Maori. A Maori woman displaying a fine catch of Northland snapper in 1910. *ARTHUR JAMES NORTHWOOD. NORTHWOOD COLLECTION, ALEXANDER TURNBULL LIBRARY, WELLINGTON, NZ. G- 6322-1/1*

CENTER LEFT: Maori made sophisticated netting from flax. This 1926 photograph shows the first steps. Peter Buck makes closed loops on the supporting strand. *ALEXANDER TURNBULL LIBRARY, WELLINGTON, NZ F-38203-1/2*

BELOW RIGHT: Coastal iwi have maintained a tradition of fishing. A Maori group fish Waipiro Bay in the Gisborne region around 1900, while a schooner unloads in the background. *F.A HARGREAVES COLLECTION, ALEXANDER TURNBULL LIBRARY, WELLINGTON, G-002608-1/1*

The catch was divided into 37 piles on the beach, and it is reported that when the catch of snapper, trevally, kahawai, tarakihi, parore, kingfish, dogfish, mackerel, and many others, had been divided equally, there were over a thousand fish in each group. If the fish averaged just a kilo each, that would be at least 37 tonnes of fish.

HOOK AND LINE (HI IKA)

Although early Maori were not unfamiliar with fishing rods (matira), fishing with a baited hook (matau) was mostly done using a handline (aho). Maori rope and cord was of high quality for the era. In the Bay of Islands in the 1830s J.S. Polack observed that 'Their fishing-lines are infinitely stronger, and fitted to bear a heavier strain, than any made from European materials.' Fibres used were mostly flax (harakeke), cabbage tree (ti) or perching astelia (kiekie). Fibre was prepared by soaking and scraping, then rolled on the thigh with constant application of saliva (a process called miri) into a single, even twine (takerekere). Multiple strands were rolled, plaited or twisted into many different forms of twine and cordage. A favoured fishing cord was the three-strand (tatoru).

The 'gorge' was used by Maori; this precursor of the fish-hook has been found at Neolithic sites all around the world. It consists of a narrow piece of bone, stone, shell or hard wood sharpened at each end and secured by a cord around the middle. The gorge is turned to lie parallel with the cord and inserted into a piece of bait. The fish is allowed to swallow the bait down and then struck with the cord, which turns the gorge sideways to wedge in the fish's throat.

The gorge is not a very efficient device however, and the more common and familiar hook type was the matau, made from bone, whale ivory, stone, shell or wood. Maori appear to have imported basic hook designs from other parts of Polynesia, then adapted them to local species and materials. In the Pacific Islands, reasonably large strong one-piece hooks could be made from pearl shell, but this material was not available in New Zealand. Small hooks could be made in one piece from bone or stone or shell, but larger hooks were often composite types with a bone, stone or shell point fitted to a wooden shank that could better withstand the greater strain.

These wooden shanks were often formed by training a tree branch into the desired shape and letting it grow and set, before harvesting it later. Tauhinu or cottonwood shrub *(Cassinia leptophylla)*, mangemange or climbing fern *(Lygodium articulatum)* and tanekaha or celery pine *(Phyllocladus trichomanoides)* are mentioned by early Europeans as being used for this purpose. Colenso, in 1869, described how branches were grown into suitable form, then heated in hot earth beneath a fire to harden the wood.

Of course there are limitations to what can be achieved with such materials. If worked into a thin, sharp point that can be easily set into a fish's mouth, these materials become too weak and likely to snap when a fish is hooked. To get around this problem early Polynesians (and other Neolithic cultures) had developed the 'circle' or recurved-point hook. Rather than penetrating the fishes' mouths by main force (as European-style 'J' hooks do), circle hooks are designed to allow the fish to hook itself, rolling into the corner of the jaw under steady pressure. Most Maori hook designs are of this type.

A composite of sketches of Onehunga, Auckland, made in July 1864 by Edward Arthur Williams shows Maori unloading fish from waka, European ships in the harbour, and a bungalow. Fish, including snapper, were a common early item of trade between Maori and Pakeha.
ALEXANDER TURNBULL LIBRARY, WELLINGTON, B-045-001

The museum curator and fish scientist Chris Paulin suggests that circle hooks made of these materials would still be too thick to penetrate a fish's flesh and instead would jam over the fish's jawbone, which tends to increase in width towards the back of the jaw. This certainly appears likely with the 'U'-shaped hooks that feature wide double or single barbs with only a narrow gap between them. They may have acted more as a jaw clip than a hook.

Sizes, patterns and construction method of matau varied widely throughout the country, reflecting the availability of different materials, fish species and sizes, and the ideas of individual hook-makers and fishermen.

To the early Europeans, with their steel 'J' hooks, traditional Maori hooks seemed 'ill-made', 'of doubtful efficacy', 'very clumsy affairs' or 'impossible looking' according to various recorders of the time. 'They live chiefly by fishing, making use … of wooden fishhooks pointed with bone, but so oddly made that a stranger is at loss to know how they can answer such a purpose,' wrote the *Resolution's* surgeon William Anderson, on Cook's third voyage in 1777.

However, a culture dependent on catching fish for survival is unlikely to persist for 700 years with a system that doesn't work. It has recently been acknowledged that modern circle hook designs are in fact superior to 'J' hooks for commercial long-lining and some recreational fishing, vindicating the Maori design.

Pre-European sinkers were made from stones. A groove was scratched around the stone to attach a line. Stone is not as dense as today's lead sinkers, and combined with the water drag and buoyancy of the line, and sometimes wooden hooks, getting down to the bottom in any sort of current must have been a problem, restricting fishing to relatively shallow water. Ledger-style rigs with wooden spreaders (pekapeka) were sometimes used when bottom-fishing.

Tamati Poata, in his 1919 booklet *The Maori as a Fisherman and his Methods*, writes of snapper: 'Almost any bait can be used for this fish, but crayfish is out and out the best. Mussels come next and then pipis.'

Elsdon Best in *Fishing Methods and Devices of the Maori* (1929) records that bait (maunu) was 'tied onto the hook with a piece of twine'. This seems likely if such soft baits were used, and indeed, some matau feature protrusions that may be 'bait notches' to help secure bait.

Poata records the technique used for fishing for snapper from a canoe: 'As you approach the fishing ground, it is wise to ascertain the set of the current. Crayfish pulp [taruru or groundbait] is then cast overboard, and while sinking, it gradually drifts towards the grounds. The fisherman then secures his bearings and anchors. The lines are quickly baited and the sport commences.'

CONSERVATION

By the time of first contact with Europeans, Maori had developed a set of practical rules to protect stocks of fish and shellfish, and to allow them to regenerate. Only certain species of fish could be taken at certain times and places. At other times a rahui or total ban on fishing could be applied to allow a heavily-fished area to recover or to protect spawning fish.

FAR LEFT: Maui hauls the North Island to the surface. It seems fitting that New Zealand's creation myth centres around fishing. *WILHELM DITTMER: (LONDON, ROUTLEDGE, 1907) ALEXANDER TURNBULL LIBRARY, WELLINGTON. PUBL-0088-049*

LEFT AND BOTTOM LEFT: These matau (fish hooks) are examples of a recurved composite (wooden shank, bone point) type (left) dating from about 1800, and an all-bone 'jaw-clip' style. *MUSEUM OF NEW ZEALAND, TE PAPA TONGAREWA MA 1054383 AND MA 1067662*

BELOW: This painting on tapa cloth by Northland artist Lester Hall reflects the importance of the species to the Polynesian/Maori culture. *LESTER HALL, WWW.LESTERHALL.COM*

There were also strict prohibitions for protecting the environment. For example, rocks rolled over had to be returned to their original position, and it was forbidden to dump waste (including human waste) or rubbish into the water.

These and other prohibitions crossed into the spiritual realms. If someone drowned, no-one could fish near the spot until Tangaroa, guardian of the sea, returned the body. The first fish taken was returned to the sea with a karakia (prayer) to bring an abundance of fish. Although today many tapu have been lost, some are still observed by both Maori and Pakeha fishermen.

EFFECT OF EUROPEANS

Fish was one of the earliest items of trade between early European explorers and local Maori, and both cultures benefited from each other's knowledge. Maori fishermen and hook-makers were not slow to see the advantages of European metals such as iron, brass and copper, and quickly discarded traditional materials. However, for a long time they preferred to make their own recurved designs out of nails or wire, in preference to the European-style barbed 'J' hooks offered in trade. Maori continued to make hooks following the circle design using modern materials well into the twentieth century.

When European 'J' hooks became freely available, it was finally accepted that they saved a lot of work in making, and they were largely accepted by Maori despite their perceived design failings. Often European hooks were altered to Maori design. Tamati Poata (1919) wrote,

'...the hooks are what are known as the European ones, but they are greatly altered by the Maori with a hammer. The bend outwards is straightened in to run parallel with the eye, the point turned upwards so as to steer itself into the fishes mouth and the barb underneath the point is bent downwards so that the fish, when hooked, will not slip off.'

The same effect was noted with cordage. European fishing lines saved much labour, and Maori began to buy them instead of making their own out of twisted flax fibre. Bought twine also began to be used for netmaking, and the kaharoa of green flax became a thing of the past.

The increasing availability of European crops and livestock reduced Maori dependence on fishing. Increasing Maori agriculture and trade in items like flax and kauri gum saw attention shift away from fishing. With the growing urbanisation of Maori, much knowledge of fishing, rituals, traditions and associated language was lost. Tribal fishing and group fishing declined from about 1885, and large communal efforts had virtually ceased by 1910, although the fishing traditions have been maintained by individuals who continue to sustain their families, and by coastal hapu.

European commercial fishing methods also began to have a huge effect on fish stocks, snapper in particular. There is little evidence that Maori seriously depleted snapper numbers, but after 200 years of European occupation the snapper biomass has been reduced by an estimated 75%.

MAORI FISHING RIGHTS

The large inshore stocks of fish that were available made it fairly easy for pre-European Maori to catch enough to sustain themselves. They had developed a sophisticated system controlling fishing rights that extended to families, villages and tribes. These dictated who was permitted to fish, when, where and how. The Maori approach was quite different to the English 'common property' approach, with its largely unregulated attitude to fishing.

The Treaty of Waitangi guaranteed Maori 'full exclusive and undisturbed possession of their ... fisheries'. The first statutory fisheries regulation was the Fish Protection Act of 1877 preserving Maori rights under the Treaty. In successive legislation, these rights were omitted in 1894 and but then reinstated in a watered-down form in 1903. The government of the day believed that the rapidly increasing pressure on fish stocks from Pakeha could be managed separately from Maori fishing. That this is not the case is obvious today, but it served to create divisions between commercial and customary fishing, and Maori and Pakeha, that still exist.

Maori fishing rights were protected under law from 1900 to 1962. On paper, by applying to the Marine Department, fishing grounds could be reserved in particular areas for meeting personal needs. However, no applications were approved in 62 years.

In 1983 a new Fisheries Act sought to bring together and update the morass of fisheries regulations that had developed since 1908. Since the fishing industry had been deregulated in 1963 and given financial incentives to develop, inshore fisheries – especially snapper – had suffered badly. It was intended that the new 1983 Act would also address this.

The 1983 Act did acknowledge Maori fishing rights, but did little to uphold them. It excluded many 'part-time' fishermen from commercial fishing, many of whom were Maori. Three years later when the Quota Management System (QMS) was introduced and harvesting rights for some fish species (based on individual catch history) were granted to licensed commercial fishermen, the excluded part-time Maori fishermen did not receive quota allocations. Maori groups challenged the QMS through the courts and the Waitangi Tribunal, claiming this was unfair because it ignored Maori fishing rights guaranteed under the Treaty of Waitangi.

Eventually, Maori customary and commercial fishing was acknowledged by the New Zealand government as an integral part of fisheries management and the interim Maori Fisheries Act of 1989 was passed. This was followed by the 1992 Settlement Act, which recognised customary rights, bought back 10% of quota already allocated, purchased half of the Sealord company and its quota for Maori and reserved 20% of all future fish quota for Maori. This resolved commercial fishing issues and ended the Waitangi Tribunal's involvement in Maori fishing claims, although the actual allocation, carried out by the government-appointed Waitangi Fisheries Commission, was bogged down for a decade trying to devise a formula that would accommodate individual claims as well as ensure that urban Maori who lived outside recognised iwi structures received some of the multi-million-dollar resource.

In 1996 a new Fisheries Act introduced and refined regulations, defining the rights and responsibilities of tangata whenua in managing customary Maori fisheries and how customary fishing could function. These include the customary 'Regulation 27', which allows special dispensation from the amateur regulations to gather seafood for hui or tangihanga under a permit system, and the appointment of kaitiaki (guardians) responsible for managing the customary take in an area.

Customary management tools include mataitai reserves – areas of traditional importance where tangata whenua are authorised by the Minister of Fisheries to manage and control the harvest of seafood for non-commercial purposes. A second type of reserve is a taiapure, a local fisheries area established over a place of special significance to tangata whenua, where a management committee can recommend general fisheries regulations for the area.

DARRYL TORCKLER

SNAPPER AND SETTLERS

03

The first mention of snapper by Europeans is found in the journals of Captain James Cook. During Cook's long voyages of discovery (his first round-the-world trip took three years), they had dried, salted meat and fish as their only preserved protein foods, so it was vital to be able to buy, trade for, or catch food regularly to restock the larder. Fishing was therefore partly for scientific investigation but mostly for supplying the crew with food.

Cook's men appear to have been poor snapper fishermen. On November 6, 1769 his ship *Endeavour* was anchored in what is now known as Cooks Bay, near the site of modern-day Whitianga. Cook recorded in his journal:

'I went to another part of the bay to haul the sean [seine net], but meet with as little success as before ... In the morning I sent the long boat to trawl in the bay, and one officer with the Marines and a party of men to cut wood and haul the sean, but neither the sean or the trawl meet with any success; but the Natives in some measure made up for this by bringing several baskets of dry'd or ready dress'd fish; altho' it was none of the best I order'd it all to be brought up in order to encourage them to trade.'

It seems that, despite the huge number of fish that must have been available in those days of near-virgin stocks, you still had to know what you were doing to put a feed on the table. By November 25, after penetrating the Firth of Thames but by-passing most of the Hauraki Gulf, Cook arrived in a large bay to the south of Whangarei Heads. There the crew of the *Endeavour* finally struck fishy gold:

'We kept standing along shore to the NW, having the main land on the one side and islands [the Hen and Chickens] on the other; our soundings were from 26 to 12 fathoms. At half past 7pm we anchor'd in a bay in 14 fathoms [about 28 m]. We had no sooner come to an anchor than we caught between 90 and 100 bream (a fish so called), this occasioned my giving this place the name of Bream Bay.'

Cook's 'bream' were of course snapper, and to this day Bream Bay, just south of Whangarei, is still a great spot to catch them. However, considering the timing – a late-November evening – it seems likely that the *Endeavour* happened upon a spawning aggregation of snapper right at that crucial change-of-light period. Given the near-virgin stock levels, it would have been difficult not to make a big catch!

LEARNING FROM THE LOCALS

Cook acknowledged the superior fishing skills of local Maori in his journal entry of 5 December, written in the Bay of Islands:

'...it affords a good anchorage and every kind of refreshment for shipping, but as this was not the season for roots [kumara] we got only fish. Some few we caught ourselves with hook and line and in the sean [seine net], but by far the greatest part we purchased from the Natives, and these of various sorts, such as sharks, stingrays, breams, mullet, mackerel and several other sorts. Their way of catching them is the same as ours, viz., with hook and line and seans,

Early European coastal settlers mixed farming and fishing for sustenance and sale (if they were near a population centre). This watercolour by John Philemon Backhouse in 1881 shows a settler's hut with fishing boats, a man carrying oars, and nets drying, in a mangrove estuary of northern New Zealand.
ALEXANDER TURNBULL LIBRARY, WELLINGTON, NZ. B-107-001

of the last they have some prodigious large made all of a strong kind of grass.'

Cook was fisherman enough to check out the sort of tackle the successful blokes were using. There were obviously plenty of fish, and later in the voyage Cook's sailors seemed to be getting the hang of catching them. Cook's Journal of 11 January 1770 records that, while near Kawhia Harbour, '... being in about 42 fathom water the people soon caught about 10 or a dozen bream'.

However, Maori were much more adept at fishing, and buying from the locals was an easy option as well as helping to establish a trading relationship. The journal of Joseph Banks, naturalist (and a man of very long sentences) on the *Endeavour*, records in March 1770:

'Every creek and corner produces abundance of fish, not only wholesome but at least as well-tasted as our fish in Europe: the ship seldom anchor'd in or indeed passd over (in light winds) any place whose bottom was such as fish resort to in general but as many were caught with hook and line as the people could eat...For the Sorts, there are Macarel of several kinds, one precisely the same as our English ones [blue mackerel] and another much like our horse macarel [jack mackerel], besides several more; these come in immence shoals and are taken by the natives in large Seines from whom we bought them at very easy rates. Besides these were many species which tho they did not at all resemble any fish that I at least have before seen, our seamen contriv'd to give names to, so that hakes [barracouta], breams [snapper], Cole fish [blue cod] were appellations familiar with us, and I must say that those who bear these names in England need not be ashamd of their namsakes in this countrey.'

FOR FOOD, NOT FIGHT

The attitude of the first European explorers to 'breams' and other fish set the scene for the future. To early European settlers, snapper, and fish in general, were a cheap and easily-caught source of protein rather than a sporting proposition. If they caught a surplus, they would sell them. Settlers who lived near a population big enough to support a market might combine farming with a little part-time fishing.

This mix continued into relatively modern times. Mavis Brambley in her 1966 book *Sea Cockies of the Manukau* records how she moved on to her husband's (then) isolated farm on the southwest Manukau Harbour in 1936 and supplemented the farm income by catching, smoking and selling snapper.

Prior to 1920 nearly all commercial snapper-fishing was done with hand-lines and set nets (mostly from small sailing boats called mullet boats), with little distinction between commercial and subsistence fishing. But as snapper were by far the most common fish caught in the north, they were probably quite important to early Europeans in this area. As an eating fish, snapper was always one of the most, if not the most preferred. It was white-fleshed fish that the European settlers were used to – not dark fish like trevally or kahawai.

The average pioneer was too busy cutting a farm out of the bush or digging for gold or kauri gum to fish solely for

LEFT: Snapper as art. This snapper was made by metal artist Miles Gordon from mild steel, coloured by heating to different temperatures with gas torch, and protected with lacquer.

MILES GORDON, REALCUTUP@PARADISE.NET.NZ

BELOW: The iconic status of snapper is reflected by its regular appearance in local art. 'Moocher' by Northland artist Lester Hall reflects the subtle colours of an 'old man snapper'.

COURTESY OF LESTER HALL, WWW.LESTERHALL.COM

On shaky ground. Auckland photographer Herman Schmidt took this shot of a thrill-seeking fisherman with his catch of snapper and kingfish in 1909.
SPECIAL COLLECTIONS, AUCKLAND CITY LIBRARIES (NZ) 31-58673

recreation: that luxury was reserved for the landed gentry and wealthy visitors. However, the publications of the late 1800s and early 1900s record these recreational fishermen as concentrating primarily on the introduced trout and, at sea, kingfish and marlin. The catch lists of the time are heavy with the titled and the wealthy; presumably they were the ones with the most leisure time (and money) to fish for sport, and snapper at that time was not considered a 'sport fish'.

One such moneyed traveller, named Kingsley, does refer to trolling a silver spoon in the late 1800s: 'You catch all kinds of fish – snappers, rock-cod and what not; the spoon is a novelty, and they rush at it like women at a new fashion in bonnets.'

Romer Grey, son of the well-known angler and writer Zane Grey, in 1930 wrote his own book, *The Fisherman Under the Southern Cross*, about a trip to New Zealand with his famous father. Romer was, however, a light-tackle enthusiast and his writings show him to have been just as excited by kingfish and kahawai as he was by marlin and mako sharks. There is only one brief mention of snapper in his work, and that as a table fish rather than a sporting proposition.

A more typical New Zealand European attitude to snapper fishing is expressed by a lighthouse keeper's son, W.E.H. Creamer, describing his father fishing at Tiritiri Matangi Island in the Hauraki Gulf around 1910:

'Fishing for my father was like what one hears about at a lying competition. His practice when fish were required was to shoot a shag, bait a hook with a piece of the flesh and toss the line into the water, off the rocks or beach. Soon he would have landed all the large schnapper he required (that was before the trawlers began).'

However, the same boy seems to have a different appreciation of snapper when referring to rock-fishing at Cuvier Island in the 1920s:

'Soon after arrival, we saw a large shape approaching through the clear water, so I lowered some strong tackle down in front of the creature. It swam up to the line, the large baited hook vanished and I jerked up onto the rocks the largest schnapper I have ever seen ... we caught two more schnapper of good size and then set off for home. I carried the one that did not get away upon my back with its head resting upon my shoulder and its tail dragging along the ground.' (*The Tall White Tower*, 1965).

SNAPPER FOR SUSTENANCE

Snapper have always been an important table fish for the New Zealanders, and the 'average bloke' of the late 1800s and early 1900s went fishing as much for a 'feed' as for fun. This became even more important during the 1920s Depression. Unemployment and poverty were widespread, and many had to fall back on their own resources. The renowned writer Frank Sargeson records surviving by 'working on a relief gang one or two days a week, and spending the rest of the week growing fruit and vegetables, fishing and writing'.

LEFT: An early beach seine in the Hauraki Gulf – a good way to harvest a feed of snapper in the days of plenty.

SPECIAL COLLECTIONS, AUCKLAND CITY LIBRARIES (NZ) 37-367

BELOW LEFT: A day's yachting on the Hauraki Gulf early in the 1900s was, as now, likely to include harvesting snapper for dinner.

SPECIAL COLLECTIONS, AUCKLAND CITY LIBRARIES (NZ) 37-90

BELOW RIGHT: The late 1950s and early 1960s saw the development of a charter fishing fleet that would take groups of amateur fishers out snapper fishing. This was particularly popular in the Marlborough Sounds and Auckland, with its large population and handy snapper grounds in the Hauraki Gulf.

ARCHIVES NEW ZEALAND/TE RUA MAHARA O TE KAWANATANGA WELLINGTON OFFICE. AAQT 6401, A26870

Many unemployed men fished off the wharves, not just to while away the time but mainly to help feed the family. Ted Bradstock (now aged 83) remembers frequenting the Miramar Wharf at Wellington as a small boy, where the main catch was snapper.

'It was before the advent of rods. These very serious amateur fishermen would each have a number of handlines out, with the top end attached to a tobacco tin full of stones that would rattle when a fish took the bait. To me as a small boy, those big fish with the silver/pink sheen, flapping on the wharf, were a wonderful sight.'

When Ted was about twelve he caught one himself:

'We had been fishing for spotties for a while when we saw this large snapper nosing around below us in about six feet (2m) of water. We baited up our kahawai line with a whole spotty and literally dangled it in front of his nose. For a while there was great suspense as he just gnawed at it but eventually he took it properly and we hauled him up in great excitement. I was very popular when I got home.'

As food freely available for the harvesting, coastal snapper and other fish were an important part of the diet to those capable of catching their own, while others took up part-time commercial fishing, or sold or bartered the excess catch. The depressed economy saw a big reduction in the operation of trawlers, partly because of running costs and partly because local markets were being supplemented by part-timers and subsistence fishermen catching their own.

A CHANGE OF ATTITUDE

Things were beginning to improve by 1935, but it took the economic activity generated by the beginning of World War II in 1939 to shake the lingering effects of the Depression. Many men who would have been of an age and inclination to fish were overseas fighting for the next five years or so, and many boats and launches were commandeered by the Navy for military work.

When the war ended, the return of servicemen from overseas saw an increase in both commercial and recreational snapper fishing. At that time Ray Doogue, through his writings in the old *Sunday Star* newspaper and a series of books, did much to popularise recreational fishing as a sport to be enjoyed for its own sake as well as putting food on the table.

In his book *Saltwater Fisherman*, he recalls being 'bitten by the fishing bug' on returning to New Zealand after a period in Germany 'as a guest of the Third Reich' (i.e. as a prisoner of war). The moment of catharsis came while drifting on a boat off the northern end of Mayor Island 'on a wonderfully shiny Saturday morning. In my mind I was congratulating myself on being alive and out of the endless fog and cold of Germany.'

Early surfcasting was highly successful and became immensely popular. A boat was not required and it could be a part of a normal family outing to the beach as well as offering the chance of a fresh fish meal. Doogue, in his book *NZ Sea Angler's Guide* recorded that during his first summer of surfcasting around 1950, '... from Christmas to the end

Up until the arrival of nylon monofilament and advances in surfcasting tackle in the 1950s, much shore fishing for snapper was done with heavy hand lines. *ARCHIVES NEW ZEALAND/TE RUA MAHARA O TE KAWANATANGA WELLINGTON OFFICE AAQT 6401, A26823*

WERE SNAPPER BIGGER IN THE OLD DAYS?

In his book *Tales of the Angler's Eldorado: New Zealand* (1926), pioneering angler Zane Grey casually mentions catching a 'forty-pound snapper ... and several yellowtail' at Sunken Reef in the Bay of Islands. Forty pounds is a nice round figure, so it seems likely that this fish was not actually weighed but 'guesstimated'. Grey was a man who didn't shrink from exaggeration; indeed, he made a very good living from writing purple prose. One critic even wrote to Grey complaining that 'If you went out with a mosquito-net to catch a mess of minnows, your story would read like Roman gladiators seining the Tigris for whales!' It's a fair guess that this snapper, the only one mentioned in Grey's landmark book, may have been a little less than 40 lb.

When trawling the old literature it seems that then, as now, a 'twenty-pounder' (9 kg) fish was a benchmark, a 30-pounder a rare trophy, and only a small handful of snapper weighing 40 pound can be substantiated.

Another early sport angler, T.E. Donne, author of *Rod Fishing in New Zealand Waters*, wrote in the 1920s that 'Schnapper have been caught as heavy as 20 lbs', and early black-and-white photographs of snapper catches do not show fish that would seem exceptional even by contemporary standards. Indeed, a trawl through the pages of any issue of a current fishing magazine such as *New Zealand Fishing News* is likely to net pictures of any number of snapper exceeding the old-fashioned twenty-pound benchmark.

The near-virgin stocks that were available to early anglers before European commercial fishing must have made catching a snapper much easier. However, present-day anglers, with the advantages of modern tackle, technology and knowledge, are probably just as or even more likely to break the twenty-pound barrier as fishermen of a century ago.

of April, I averaged eight snapper plus a good number of kahawai, gurnard and other fish for each visit to the beach.' He also reports that 'at … Ninety-Mile Beach … it is recorded [that] one angler landed forty-six snapper aggregating 484lb, in a day's fishing.' Such catches were probably the result of snapper stocks replenishing during the Depression and Second World War, combined with the tapping into the relatively untouched surf zone beyond the range of handlines, but inside the trawl zone. The post-war increase in recreational fishing and the revival of commercial fishing soon saw catches drop to lower levels.

It was also around the late 1940s and early 1950s that increasing numbers of boating and fishing clubs began popping up around the country. Clubs such as the Ninety Mile Beach Rod and Line Club, the Pania Surfcasting Club in Napier and the Auckland Outboard Boating Club were all formed by keen Kiwis who wanted to be able to get on the water and enjoy a spot of fishing, but at little expense. Many early clubs were surfcasting clubs, and the New Zealand Angling Association (forerunner of the current New Zealand Angling and Casting Association) held its first-ever fishing carnival at Ninety Mile Beach in 1958. About 200 anglers attended and over 24 hours landed 1800 lbs of snapper.

TACKLE DEVELOPMENTS

With snapper fishing mostly regarded as a food-harvesting exercise prior to the Second World War, amateur fishermen largely copied the commercial techniques of the time. Handlines, nets and longlines were standard equipment. As late as the 1960s many of the Baby Boomers first learned to fish with a handline before advancing to a rod and reel.

At that time it was still possible to see the handline beach fishing methods of the earlier era being used: the fisherman would uncoil the cord line from the stick it was wound around, laying it out carefully on the beach, then swing the sinker and line around his head, building momentum before finally launching the rig out to sea with a flourish like a discus thrower. The line was propped up on a long stick, the end secured around a heavy log, and he would sit back and wait for a bite.

Aided by the technological advances spurred by the Second World War, fishing tackle became much easier to use. Synthetic lines, first Dacron and then nylon monofilament, replaced cotton and gut. DuPont had announced the invention of nylon in 1938, and started commercially producing primitive nylon monofilament fishing lines before nylon production was requisitioned to make parachute fabric, army bootlaces and other materials for the war effort. It was not until 1958 that DuPont produced a reasonable quality nylon monofilament fishing line, the original Stren, which saw nylon lines become popular.

In the 1950s, recreational surfcasting took off in a big way. Unlike the well-designed eggbeater-style reels available today, it took skill to cast those early reels without getting a backlash or 'bird's nest'. Temperamental early tackle, such as the Nottingham-type centrepin reel, was replaced by the Malloch (side-cast), early threadline (eggbeater) types and overheads such as the Pfleuger Sea King.

Even in the early 1970s tackle stores still had big racks of Rangoon canes, 16 to 18 feet (5.3–5.9 m) long on their back walls. Any downtime the staff had was spent binding guides on to the canes, or customers had the option to buy a bare cane (either in one piece or with a centre join so it could be broken down) and customise it themselves. A little later fibreglass rods, first solid then hollow, came on the scene, and the cane rods went the way of the dinosaur.

CHARTER FISHING

The late 1950s and early 1960s also saw the development of a charter fishing fleet that would take groups out for a half-day or day of snapper fishing. This type of fishing was particularly popular in Auckland, with its large population and handy snapper grounds in the Hauraki Gulf. Tackle was simple – a handline, and a sack tied to the rail where the angler stood, to put the catch in. A more refined version of this charter fishing tradition continues today, and snapper charters are popular in most ports in the North Island and Marlborough Sounds.

Advances in fishing tackle and early high success rates saw surfcasting become hugely popular in the 1950s.
ARCHIVES NEW ZEALAND/TE RUA MAHARA O TE KAWANATANGA WELLINGTON OFFICE AAQT 6401, A48888

DARRYL TORCKLER

COMMERCIAL FISHING FOR SNAPPER

04

European commercial fishing really got underway in 1876 when the Marine Department was established; licensing was introduced in 1894 and vessel registration in 1903.

Albert Sanford, founder of the fishing company that today still bears his name, fitted the *Minnie Casey* with a beam trawl for fishing in the Hauraki Gulf in 1899. By 1902 there was massive opposition as trawling was thought (wrongly) to kill the spawn of snapper on the bottom of the sea and destroy their food (habitat damage was more likely). A 'no-trawling' line was drawn in the Gulf, forcing the *Minnie Casey* out of business. Trawling did not resume in the Gulf until 1915, at which time the more efficient otter trawling (with boards or doors) was used.

In 1901 and 1908 the Napier trawler *Doto* was chartered by the Marine Department to search for new fishing grounds around the country. Snapper were among the most plentiful fish, especially in the north, although the Marine Department was already noting local depletion of some grounds close to Auckland. By 1917 the national trawl fleet had grown to 70 steam or diesel-powered trawlers and by 1920 it had become necessary for Auckland trawlers to venture as far afield as the Bay of Plenty and around North Cape to the waters off Ninety Mile Beach to obtain satisfactory catches.

SELLING THE CATCH

Throughout northern New Zealand snapper was almost everyone's preferred fish. It was sold fresh or smoked, as fresh fillets or whole fish. Snapper was the usual fish on hotel and restaurant menus and in fish and chips.

In the 1860s Albert Sanford started a business that would eventually becomes New Zealand's largest and only publicly-owned fishing company, Sanford Ltd. He caught and smoked snapper from his home on Rakino Island, selling them in Auckland. Gradually he built up his business as a merchant and retailer until he was able to buy his own trawlers in 1915. By 1917 Auckland and Thames were yielding more than half the country's fish catch, much of it snapper that Sanford sold through his 20 shops and two restaurants in Auckland.

Trawling became banned in the Hauraki Gulf to conserve fish stocks and in 1924 Sanfords helped pioneer Danish seining, which soon became the main snapper-fishing method. By this time fisheries inspectors had already sounded the first warnings that snapper were becoming over-fished, especially in the Hauraki Gulf, and by 1929 a raft of seining restrictions were in place.

Although Sanfords dominated the northern fishery, with factories in Auckland, Thames and Tauranga, many Dalmatian emigrants also took part. In response to laws discriminating against 'aliens' after World War I a number of them formed a company like a trade guild, which in 1936 became the Auckland Seine Boat Association Ltd. The other main companies catching snapper by trawl and seine were Auckland Fisheries and Kia Ora Fish Market. From Napier, the New Zealand Trawling Company supplied snapper to the lower North Island, and a small fleet in Tasman Bay caught snapper near the southernmost limits of their distribution.

LEFT: In the days before fish shops, fish were often sold at the wharf. This 1905 shot of the Auckland waterfront and ferry terminal shows snapper being sold direct from the boat at lower left.
SPECIAL COLLECTIONS, AUCKLAND CITY LIBRARIES (NZ), 1-W1774B

BELOW: By 1917 Auckland and Thames were yielding more than half the country's fish catch, much of it snapper that Sanford Ltd sold through its 20 shops and two restaurants in Auckland.
SPECIAL COLLECTIONS, AUCKLAND CITY LIBRARIES (NZ), 1-W1774B

BELOW LEFT: Commercial fishing used to be very hands-on and labour intensive. Here fishermen in the 1930s at Island Bay, Wellington, load their catch from the beach on to an open truck for transport to market.
ALEXANDER TURNBULL LIBRARY, WELLINGTON, NZ. 1/2-047912-G

By the outbreak of World War 2 the industry was well established and snapper was its mainstay, with 7000 tonnes a year making up nearly 40% of the total catch. Tarakihi and gurnard were the main by-catch.

New Zealand had first exported small amounts of fish as early as 1883, but imported much more preserved fish. In 1912 frozen wetfish exports, all to Australia, exceeded £34,000 and by 1928 the value had grown to nearly £75,000. When snapper were first exported to Australia in 1933, blue cod was the main export fish.

TOUGH TIMES FOR INDUSTRY

As New Zealand recovered from the Depression, rigorous controls were placed on most industries. The 1936 Industrial Efficiency Act strictly dictated the price of fish and the number of commercial fishing licenses. Efficiency was discouraged and fishing was regulated to a degree that seems unimaginable today. Fishing boats were only permitted to land the catch at their own base port, a restriction that was intended to safeguard local supplies but only made the economics of fishing more difficult.

The Sea Fisheries Investigation Committee in 1937 reported that 'the best interests of New Zealand would be served if the fish export trade were almost totally dispensed with' and in 1956 the Caucus Fisheries Committee said it expected exports to dwindle as local demand increased with a rising population: 'In time, the export of fish from New Zealand may cease altogether.'

The Marine Department then had only two fisheries administration staff and nobody fished deeper than about 90 metres, well within the maximum depth range of snapper.

By 1950 the need to protect snapper had driven the government to introduce further restrictions on Danish seining, and trawling in the Hauraki Gulf was banned from October through February.

Export markets were slow to develop and New Zealand was losing out to more enterprising countries which vigorously marketed their fish. South Africa, Japan and Britain produced better-quality fish, and their industries were better developed as they had incentives and subsidies that were not yet available to their New Zealand counterparts.

Australia was desperately short of fish and New Zealand would obviously be a better trading partner if only it would develop its fisheries and raise its quality control and processing standards. Our snapper were not the premium fish they are today, and as late as 1962 New Zealand still had less than a fifth share of the Australian market.

TAKING OFF THE BRAKES

Since the 1936 Industrial Efficiency Act fishing had been in stagnation: it was virtually impossible to build a new boat and get a fishing license. Twenty-four years passed before a special Industrial Development Conference provided a call to enlarge the trawling industry, and in 1962 a parliamentary select committee investigated how to do this. Its report, known as the Scott Report after the committee's chairman and first Minister of Marine, W.J. Scott, became the

FISHING METHODS FOR SNAPPER

Longlining: baited hooks used to catch top-quality snapper, particularly for export. Highly labour intensive but can be done from a small boat.

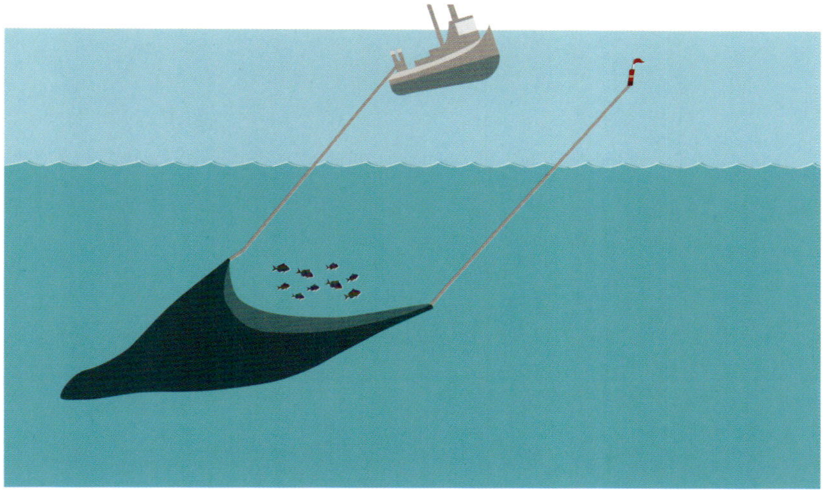

Danish seining: similar to trawling but the vessel stops after laying out the net and tow ropes. The ropes are then winched in, herding the fish into the net. This is an efficient and selective method of catching snapper, and does not crush the fish the way trawling does.

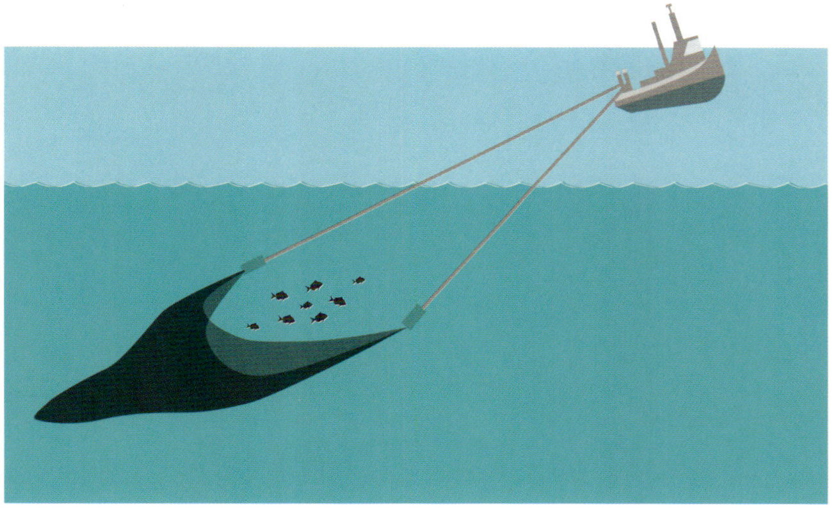

Trawling: towing a net along the bottom with wire ropes attached to boards ('doors') to spread the net open. Catches a mixture of bottom-dwelling fish including tarakihi and gurnard as well as snapper, but the fish tend to become bruised and crushed in the pocket ('cod end') of the net. Requires a comparatively large and powerful boat.

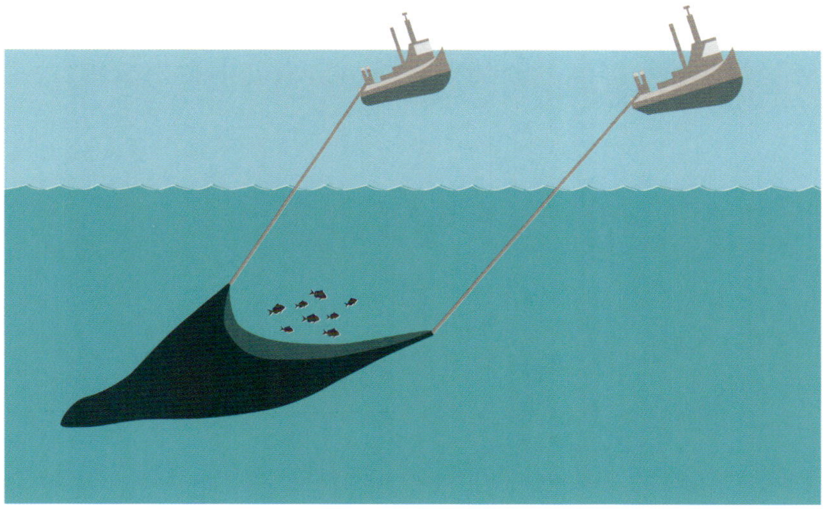

Pair trawling: two vessels pulling one side each of a large net. No trawl boards are required. A highly efficient method that sweeps up virtually every fish in the net's path.

founding document of the modern industry. This report recommended freeing up access and providing tax incentives to encourage fish exports. It also established the Fishing Industry Board (FIB) to promote the industry. Nothing much was said about recreational fishing or conservation.

Suddenly in 1963 the industry was opened up to all comers. In 1964, 136 new vessels began fishing; 209 more joined them in 1965. By 1967 investment in the NZ fishing fleet had almost doubled to £8.5 million. Many new boats had electronic equipment which made fishing safer and easier. With the exclusion of the Japanese snapper fleet from New Zealand waters around this time, the export market for 'green' snapper burgeoned.

Wherever snapper occurred they were the number-one target species and no other coastal fish was so rapidly and greatly depleted. In the early 1970s the advent of pair trawling (two boats towing one extra-large net between them) made it easy to sweep up all the fish over large areas of sea bed. Although there were numerous regulations and restrictions, they were poorly enforced and ineffective. The real need was somehow to limit the catch, but this could not be done while more and bigger boats were going fishing. It was a case of too few controls, too late.

SNAPPER WAS KING

Snapper fishing was done by a diverse mixture of company-owned boats with crews who were paid a percentage of the catch value (share fishermen) and owner/operators who often had a large personal mortgage on their boat. Generally the company boats were larger and more powerful, so they could tow trawl or Danish seine nets. Other owner/operators kept their overheads low by fishing with set nets or longlines from small converted launches which did not require an expensive annual marine survey because they were less than seven metres long. Some even used dinghies, and many only fished part-time or seasonally.

Between 1964 and 1971 the total New Zealand fish catch increased by a half, the volume of exports nearly doubled, and the value increased fourfold. Snapper was king: in 1971 it made up a third of the total wetfish catch and was worth $2.2 million.

The reported trawl catch of snapper rose from 5100 tonnes in 1962 to more than 7800 tonnes in 1967. Between 1968 and 1969 the reported Danish seine catch almost doubled, to 3400 tonnes. Longline and set net catches remained fairly steady during this period at around 1500 tonnes per year. Actual catch was probably rather higher. The Hauraki Gulf furnished a third to a half of the total catch.

THE GOLD RUSH

By the early 1970s far more snapper was being exported than sold locally, and the government was doing all it could to encourage this. There were a range of export incentives, including loans and tax breaks for buying vessels and building factories, chillers, freezers and ice-making plants. Building boats in New Zealand was expensive, so some

Red sea bream (snapper) have long been important to the Japanese. Ichiryusai Hiroshige (1797–1858) was one of the last great Japanese ukiyo-e (print) artists. Here he portrays a madai or red tai (red sea bream) with sprigs of sansho (Japanese pepper), from a set of fish prints published from 1832–4. The accompanying text is translated as *The red bream glows in the sea like a small sunset over water*.

SNAPPER AND JAPAN

The Japanese built up a distant-waters fishing fleet as part of their post-war redevelopment. When they began exploratory fishing around New Zealand in 1956 they discovered that snapper were practically identical to tai or madai, a species that was greatly valued back home. It was simply *the* fish to serve on ceremonial occasions.

Japanese long-liners and trawlers flocked to NZ's northern coasts. Their long-line operations consisted of a mother ship with smaller dories that worked close inshore. The fish were more plentiful closer in. Before long, infringements of the territorial sea, which extended only three miles from shore, were being regularly reported. The Japanese were taking 'our' fish and anti-Japanese feelings still ran strong – it was barely 20 years since World War 2. In 1964 Japanese longliners caught 5800 tons of New Zealand snapper.

The government could not ignore this, but at the same time was cautious about antagonising an important trade partner. In 1965, a nine-mile fishing zone was declared outside the three-mile territorial sea, and the Japanese were gradually forced to fish outside what became known as the '12-mile limit' by the end of 1970. In most areas there simply weren't enough snapper that far out, and more infringements followed with fishing dories being sent well inshore of the mother ship. Japanese boats were sometimes encountered by New Zealand commercial and recreational fishermen.

Finally the Japanese snapper fleet packed up and went home for the last time in the early 1970s, leaving a legacy of circle hooks and the net-encased glass longline floats that are now classed as 'Kiwiana' at batches around the country. But they had also created a market for New Zealand snapper in Japan, and exports there by local companies took off.

OPPOSITE, CLOCKWISE FROM TOP LEFT: A longline set in 20 metres of water off Rockells Bay, Northland, by Adrian Douglass of the *Emma Jane* in 1985. The backbone (main line) and nylon trace on a running 'Sea Harvester' clip can be seen. *MIKE BRADSTOCK*

John Walker of Leigh lifts a seine net of snapper on to the *Melodeon* in the Hauraki Gulf. *KIRK HARGREAVES*

The late 1970s saw an upsurge in longline fishing for snapper, partly driven by rising fuel costs. It was less capital-intensive, used far less fuel than trawling and produced higher-value fish. Tutukaka-based longliner Cushla Amos unhooks a prime fish on the *Jenny B*. *KIRK HARGREAVES*

Longline-caught snapper are never handled with bare hands as the heat will discolour the skin. They are immediately killed with a spike to the brain and put into a slurry of flake ice and seawater, preserving the colour and arresting deterioration. Later the chilled fish are transferred into fish bins and neatly arranged like sardines in a can. *KIRK HARGREAVES*

special import licenses were issued to bring in fishing vessels from overseas. There were even subsidised fishing vessel ownership savings accounts to help encourage young skippers to get their first boat.

The domestic market only grew slowly. Promotion of fresh fish only began after the Fishing Industry Board was established in 1964. In 1967–68 the FIB launched a television advertising campaign based around the slogan 'Fish – the family dish', complete with recipe leaflets, and commissioned cooking personality Alison Holst to do demonstrations featuring fish-based dishes. Fish was sold mainly fresh from shops. For a long time takeaways usually meant fish and chips; hamburgers did not really catch on before the 1970s.

By the late 1960s the word 'over-fishing' was beginning to be heard among recreational and commercial fishermen alike. There can be no doubt that snapper were becoming fewer and on average smaller; yet the official position at the 1968 National Development Conference was still that catch rates were continuing to improve and there was 'no evidence of serious over-exploitation'. Ten years later, however, when the snapper catch reached an all-time high of 18,000 tonnes, there could be no doubt of it. This figure is not far short of twice the yield the Ministry of Fisheries today considers acceptable.

The late 1970s saw an upsurge in longline fishing for snapper, partly driven by rising fuel costs. It was less capital-intensive, used far less fuel than trawling, and produced higher-value fish. Snapper caught were immediately killed with a spike to the brain and placed in a slurry of flake ice and seawater which rapidly chilled them down, preserving colour and quality. Later the chilled fish would be transferred into fish bins and laid upright, arranged nose to tail like giant sardines in a plastic can, and covered with a lid. Great care was taken not to dislodge scales or mark the fish. In the factory these top-quality fish (known as 'iki' snapper after the Japanese term 'iki-jima', a technique of killing fish with a spike to the brain), were swiftly packed in polystyrene bins with small bags of ice alongside, and air-freighted the same day.

In recent times Japan began large-scale farming of snapper, and virtually all NZ exports of snapper to Japan ceased. Other markets were developed for these premium fish, particularly in Hong Kong, China and Taiwan. This is still the main way snapper are caught and processed, although about a third of the catch is taken by trawl and not used this way.

UNDER PRESSURE

Throughout the 1970s MAF had been under increasing pressure to acknowledge the overfishing problem and do something about it. Although the bureaucrats were hamstrung by weak legislation, they could have done more had there been the political will. Part of the problem was that the first Minister of Fisheries, Duncan Macintyre, appointed in 1976, held the marginal Bay of Plenty seat, an electorate where fishing was simply too much of a nettle to grasp. His 'Think Big' fishing policy had the unintended effect of increasing pressure on an already stressed snapper fishery.

MV MELODE

LEIGH FISHERIES

This was seen at its worst in Tasman Bay (and reflected in many other snapper fisheries) where the last of the spawning schools of snapper were cynically cleaned out in 1978-79 by six large new trawlers whose owners had received special dispensation to import them duty free, subject to the condition that they were to fish in deep water. The spectacle of Skeggs' and Sealords' 40-metre behemoths pair-trawling close to shore enraged recreational anglers, and to make matters worse MAF then responded to the shortage of inshore fish around Nelson by revoking the minimum size on snapper and tarakihi. One fish processor, asked what was the smallest size of snapper he would accept for processing, responded, 'If you can see it I'll take it.' This travesty saw the snapper catch in this area, SNA7, collapse, falling from 1776 tonnes in 1979 to less than 300 by 1985. The permitted catch in 1988–89 was 372 tonnes but less than a half of that could actually be caught.

THE QUOTA SYSTEM

MAF experimented with new tighter management options including 'controlled' fisheries and an unwieldy bureaucratic process called fishery management planning, but these were supplanted in 1983 by the decision to implement a quota system for inshore fish, as had already been done with deep-sea fisheries. Snapper were among the first inshore fish allocated, in 1986. A total allowable catch of 6546 tonnes was set and divided up among commercial fishermen according to their catch history over the previous three years.

This established property rights: each tonne of snapper quota conferred on its owner the right to catch a tonne of snapper every year in perpetuity. Effectively the fishery became owned by the people who owned the boats and caught the fish. However, quota could be leased, sold, or bought, and the big fishing companies set out to buy up all the quota they could get, to ensure security of access to fish in the future.

In 1987, a tonne of snapper quota in the main snapper fishery, SNA1, was worth about $13,000 but this rapidly increased to nearly $57,000 by the season of 1993–94. When first handed out, snapper quota was widely spread among 752 quota holders. Many fishermen simply could not resist the offer of a large cheque, and by 1996 there were only 341 holders of snapper quota and 63% of the quota was owned by just six big companies. Some of the fishermen who had sold out invested the money wisely, but many are the stories of fishermen who frittered it away.

Aggregation of fish quota by corporate owners in the late 1980s set in motion a trend away from owner-operated vessels to company-owned vessels which paid the crew a percentage of the catch value. Some snapper longline fishermen and a few trawler operators continued fishing by leasing quota from companies, subject to the obligation to sell to them at a price dictated by the company. For the most part the fishermen surrendered their power to own and control the fishery. The market became flooded with second-hand boats which were of little value when offered for sale without quota to catch. There were winners and losers and much bitterness over the outcomes.

Commercial snapper landings for the SNA1 fishery

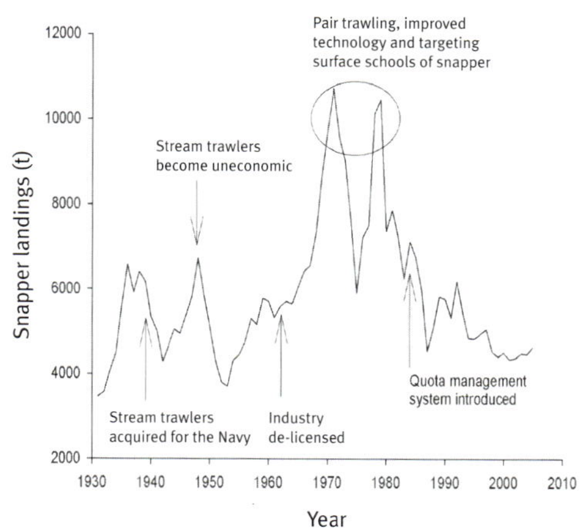

ABOVE: Cause/effect changes in commercial catch rate of snapper in the SNA1 region over time, reflecting national trends. Catch data from pages 722–758 of the Fisheries Assessment Plenary, May 2006: stock assessments and yield estimates.

MINISTRY OF FISHERIES SCIENCE GROUP

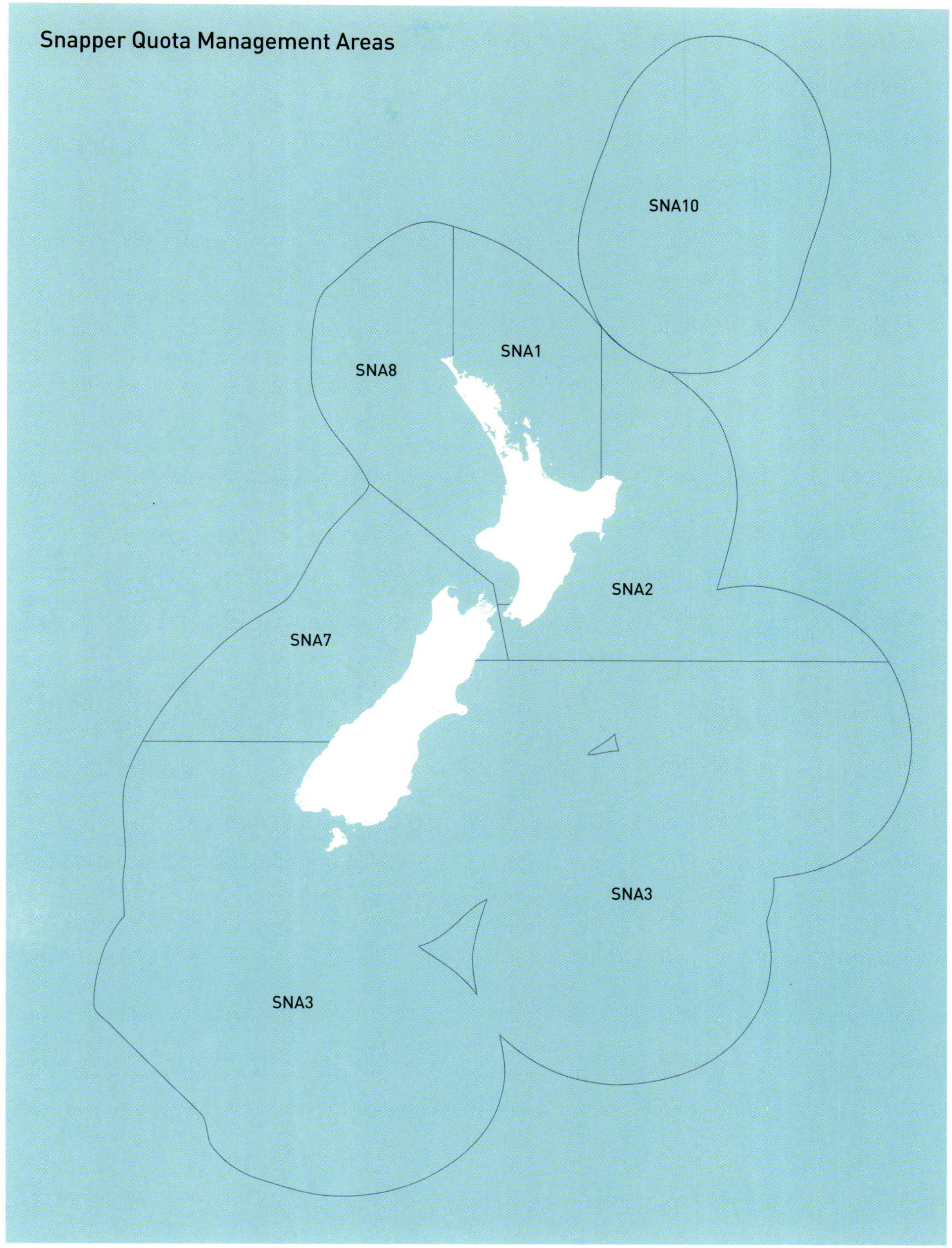

UNDERMINED

Unfortunately the snapper quota, initially set at levels which would allow the badly-depleted fishery to rebuild, was undermined by two things: the Quota Appeals Authority, and the 'deemed value' scheme.

The Quota Appeals Authority was set up to hear fishermen's individual grievances over the amount of quota they had been awarded. According to some who were there at the time, it became a bit of a scandal, with fishermen pleading poverty, being granted extra quota, then selling it for cash. In addition to the original 6546 tonnes of snapper quota allocated, the QAA dished out a further 1679 tonnes and the formal Total Allowable Catch was adjusted upwards to allow for this.

The 'deemed value' system was brought in to encourage commercial boats to land fish that they caught accidentally (or 'accidentally-on-purpose') but did not hold quota for, so these fish were included in catch reporting, and not just discarded at sea. Depending on your point of view, deemed value can be like a well-deserved penalty for catching fish that should have been avoided, or it can be an unfair penalty for accidentally catching certain kinds of fish that could not be avoided while legitimately trying to catch other fish. And there is another complication: sometimes the deemed value is low and the fish price is high, meaning there is a profit to be made by 'accidentally' catching fish for which no quota is held. This happened with snapper, and the effect was to take even more fish, over and above the level originally set to allow a rebuilding of the fishery and the 'extra' doled out by the QAA.

PART-TIMERS EXCLUDED

When the Fisheries Act was revised in 1983 so-called 'part-time' fishermen suddenly became excluded from the industry. This was a catastrophe in rural communities where many Maori seasonal workers depended on fishing for their income during part of the year (see Chapter 2). An important part of this was snapper catches, taken mostly by longline and set net from small boats. MAF applied the simple rule that anyone earning less than 80% of their income from fishing was a part-timer, failing to draw a fair distinction between part-time and seasonal fishing. This was a major factor in triggering off Maori claims under the Treaty of Waitangi.

With the removal of part-timers, over-fishing making it harder to earn a living, and a moratorium on the issue of new licenses in 1981, much fishing was consolidated into larger and more efficient vessels. However this effect was less pronounced in the snapper fishery, where the best economic return was to be had by longlining, which could be conducted from small boats.

SNAPPER TODAY

Since the era of worst overfishing in the 1970s, stock regeneration has been slow. In 2007 for example the total

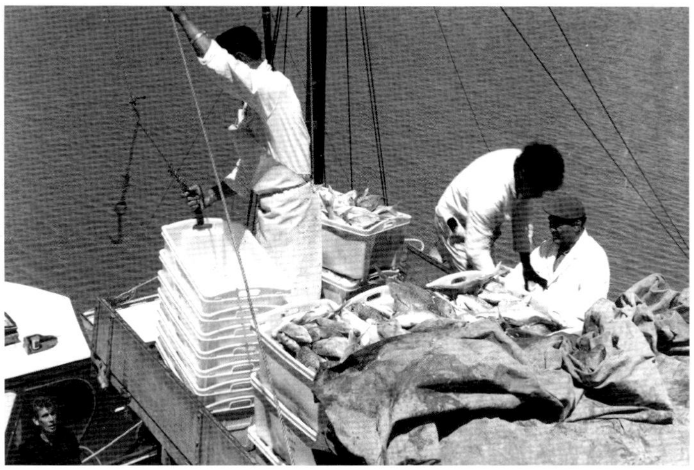

FAR LEFT: Unloading a catch of snapper from the prolific Hauraki Gulf fishery in Auckland in 1970.
NIWA LIBRARY

LEFT: Genial fish shop owners Antonio Cuccurullo and Camillo Amitrano sold fish at Wellington Fisheries Ltd in Cuba Street, Wellington, for 36 years before retiring in 1990.
DOMINION POST COLLECTION, ALEXANDER TURNBULL LIBRARY, WELLINGTON, NZ. EP/1990/1921/7A-F

BOTTOM: Skeggs' huge pair trawlers *Hawea* and *Wanaka* were part of the massive overfishing which saw spawning snapper nearly wiped out in Tasman Bay by the early 1980s.
NELSON PROVINCIAL MUSEUM, NELSON MAIL COLLECTION

snapper allocation was 10,091 tonnes, of which 31% was reserved for recreational and traditional fishing. This left 6357 tonnes for the commercial fishery.

More than two-thirds of the commercial catch is exported: 4178 tonnes valued at $29 million in 2006. This is less than 1% of the weight and barely 2.5% of the value of the total fish catch ($1.1 billion, 517,000 tonnes), and reflects the massive growth of deep-sea fishing for species such as hoki and squid which scarcely used to figure at all in fish catches.

SNAPPER AQUACULTURE IN NEW ZEALAND

Currently there is no commercial production of farmed snapper in New Zealand and no research is being conducted on the aquaculture of snapper. Farming snapper to marketable size takes years and is expensive. The market price for snapper is set by availability from wild stocks, and while snapper are still abundant this price is too low to justify the expense of farming them. In Japan wild stocks are low and the domestic demand for snapper is met by aquaculture and re-seeding. In New Zealand, the most recent snapper aquaculture ventures were at Pah Farm at Kawau Island, where Moana Pacific was farming snapper for commercial production, and in Nelson where excess snapper grown for food quality research were released into Tasman Bay.

THE FUTURE

In 2007 MoF issued proposals confirming that customary and recreational fishing should have at least 20% of any stock and receive priority over the commercial sector. It is likely that there will be adjustments to the total allowable catch as stocks fluctuate or as better data on the fish abundance is known. It is also likely that the carve-up between commercial and non-commercial uses will change.

The last time the biomass of each snapper stock was assessed, all areas were thought to be at an unsustainable level (except for the east Northland sub-stock of SNA1 where the biomass was said to be around the level of sustainability). The projections for all areas did predict, however, that snapper biomass would surpass the level of sustainability in the next 20 years.

BEFORE THE COURTS

Over the years, attempts to reduce commercial catch in shared inshore fisheries through the quota system have often been aggressively resisted by commercial interests. An example is the 1996 case between Sanfords and the Ministry of Fisheries. In 1995, the Ministry attempted to reduce the amount of snapper taken in SNA1 to help the fishery recover. Recreational fishermen's daily limits were reduced substantially, from 20 snapper per person per day down to nine in SNA1.

Recreational anglers both accepted and welcomed the cuts for the sake of the future of the fishery, but commercial operators refused to accept any cuts and took the matter to court. Initially the Ministry won the right to manage the fishery, but the decision was overturned on appeal for what was considered by many to be a technicality, and a commercial reduction was not imposed.

This case considerably soured relationships with recreational fishers, who were from then on reluctant to take any further cuts, or to trust the fishing companies to do their share in rebuilding the fishery.

Finally, in 2006, in response to the Ministry of Fisheries' efforts to cap the catches of recreational anglers (which would result in continuing reductions of individual catch limits as angler numbers increased) in the 'Soundings' proposal, recreational anglers went to the High Court to establish their rights in law. Known as the 'Kahawai Legal Challenge' the case, although based on the setting of kahawai quota levels, was really about establishing a legal precedent that applied to all shared fisheries, including snapper. It was led by on-line public fishing rights group option4, the NZ Big Game Fishing Council, the Recreational Fishing Council and Te Runanga A Iwi O Ngapuhi. Although the case was between recreational anglers and the Minister of Fisheries, the commercial fishing industry could see that case law in favour of recreational anglers might potentially affect their profits, and were quick to get involved.

In a landmark decision recreational fishing was recognised as a right that stems from common law; allowing for people's wellbeing (health or physical welfare) is the starting point when setting the Total Allowable Commercial Catch; people provide for their wellbeing by catching and purchasing fish; and both customary Maori and recreational fishing interests must be provided for where they exist. The same does not apply for commercial interests.

This, for the first time, established the recreational fishing right in law, and placed it ahead of commercial fishing. The Minister of Fisheries and the Ministry accepted the decision but, predictably, the commercial industry appealed.

DARRYL TORCKLER

THE NORTHEAST COAST

05

The coast running north from the aptly-named Bream Bay to the top of the North Island is wildly beautiful, with rugged coastlines, numerous islands, rich estuaries and magnificent golden sandy beaches – all fabulous country for snapper.

The Whangarei Harbour area is quite different from the estuarine areas further north. There are few snapper spawning aggregations inside the harbour and most of the early-season activity is out in Bream Bay and around the Hen and Chickens Islands. In early spring there is a brief run of big fish through the harbour, and this is matched by a similar run of good-sized fish on the adjacent coast. Typically the harbour doesn't start to fish well again for snapper until midsummer. From early autumn through to midwinter it is at its best, with plenty of good fish in areas with strong currents near the harbour mouth.

The Hen and Chickens Islands are excellent for snapper fishing, with big 'kelpie' snapper in the shallows, spawning schools in spring and early summer, and plenty of big fish in the medium-depth grounds. It is not uncommon to catch both hapuku and big snapper off the same reef at the 'Hen and Chicks'. Deep drifting with large baits is a popular technique.

Bream Bay has many sandy areas and a fair amount of low-lying reefy ground. Many people fish over these reefs, and drifting and jigging with plastic lures is very productive. The southern end has a small area of shallow water and reef good for 'straylining' with light tackle.

Straylining is fishing with unweighted or lightly weighted baits in shallow water, using just enough lead to counteract the current so the bait can get down to the depths. Snapper are not always right on the bottom so it is important for the angler to be attuned to the tackle, sensing subtle irregularities that may mean a fish has taken the bait.

The biggest fish are often deeper, being shy of approaching the noise, shadow and vibration of a boat in shallow water. A short cast down-current often produces better results than dropping the line straight over the side. When the fish bites, the angler should let it have a little line, then throw the reel into gear, hopefully hooking the fish.

From Bream Head to Cape Brett the extensive rocky coastline is interspersed with small harbours, river systems and a few sandy beaches. A brief run of big fish occurs from late August through to late October. Otherwise, it fishes best from March till late June for snapper of all sizes. There are few spawning schools in the warm season and limited nursery waters. The Poor Knights Islands have good spawning aggregations, but are now a marine reserve.

The jagged coastline in this section of Northland is a rock-hopper's and strayliner's paradise. Most of the best fishing occurs around hard reef structure in relatively shallow water.

From the Bay of Islands to Whangaroa Harbour, big spawning schools reflect the huge area of nursery waters available and the large, gently-sloping underwater plain that extends well out to sea. This vast shelf is less than two hundred metres deep and is all potential snapper territory.

The Bay of Islands is the central spawning ground for a much wider snapper habitat. Baitfish activity is often

LEFT. Folklore has it that in the north, when the pohutukawa trees are in flower, the snapper will be on the bite.
SAM MOSSMAN

CENTER LEFT: Large snapper are significant reef predators. The author took this 28 lb fish on a livebait near Matai Bay.
IVAN LESLIE

FAR LEFT: Big old snapper like foul ground. This 25 lb fish was taken by Rick Wakelin at Waiwiri Rock, just south of Cape Brett.
SAM MOSSMAN

BELOW: The massive shellfish beds of Snake Bank in the Whangarei Harbour are a huge food resource for snapper.
SAM MOSSMAN

intense and the local snapper frequently prey on the anchovy and pilchard schools. In autumn and winter reasonable numbers of snapper hold over in shallow water in harbour channels and over shallow reefs.

The stretch of coast from Whangaroa to Doubtless Bay is a wild rocky coastline with big snapper in reefy environments. In the harbour and around the many islands there are good nursery waters and the spawning schools are surprisingly large. The autumn snapper fishery in the shallows of the fiord-like Whangaroa Harbour is worth travelling for.

Doubtless Bay is another strong spawning area which almost matches the Bay of Islands in terms of productivity. Massive nursery areas in shallow waters spread out behind extensive reef systems in the Bay.

From the hill-tops Doubtless Bay looks quite sandy, but it is actually filled with extensive areas of rocky bottom. Big shell-lined channels lie between the reefs, rich in scallops. These are prime snapper waters and fish very well almost year round.

The Karikari Peninsula is a wild rocky coastline that sees an influx of boats during the summer season. It is a very good fishery in autumn, winter and early spring, but can suffer from a deep-water upwelling in late spring and summer. The coastal waters drop away quickly to one hundred metres or more deep, and westerly or southerly winds will pull cold water to the surface, depressing spawning activity. Just around the corner in Karikari Bay there is shallower spawning water that fishes better at this time of year.

The area from Karikari Bay to North Cape includes Houhora Harbour, Great Exhibition Bay and Parengarenga Harbour. This massive area is one of the more exposed parts of Northland. Harbours are few (Houhora, Rangaunu, and Parengarenga) but large. A gently-sloping continental shelf from Henderson Bay right through to Parengarenga Harbour offers a vast expanse of excellent snapper habitat.

There is a lot of commercial snapper fishing done in this area. The exposed water suits the big longline boats that now dominate the fishery. Despite this, recreational fishers still find plenty of snapper. The fishing for large snapper is probably better here than anywhere else in New Zealand. Some of the reef structures near the mouths of harbours yield large numbers of big snapper year after year.

East Beach, Henderson Bay, and Rarawa Beach are all notable for producing big snapper in the surf, especially after a gentle easterly breeze, from late summer onwards. Rock fishing is excellent but access can be difficult. For a short time in late autumn, good snapper invade the upper harbour and can be found in remarkably shallow water well up into the estuaries. Both Houhora and Parengarenga harbours are noted for very clear water.

THE SCHOOLING SEASON

The annual influx of spawning snapper into the schooling grounds of Northland's north-east coast typically begins sometime in late October. Snapper are serial spawners; they will spawn a number of times, given favourable

LEFT: Extensive areas of foul and large mid-depth bays provide ideal snapper habitat in the north. This area, south of the Bay of Islands, is a good example. *GRANT DIXON*

BELOW: Snapper are aggressive feeders and can be caught with a wide range of techniques, including saltwater fly-fishing. *CRAIG WORTHINGTON*

BOTTOM LEFT: Six- and seven-metre-class charter boats, such as this Bream Bay operation, are popular with anglers, who not only have the use of the boat, but importantly the local knowledge of the skipper. *SAM MOSSMAN*

BOTTOM CENTER: The northeast coast of the North Island has a wealth of excellent snapper habitat. Josh Worthington with a prime fish caught from the rocks in ideal conditions. *CRAIG WORTHINGTON*

BOTTOM RIGHT: Snapper as TV stars. Cameraman Kerran Packer films host Matt Watson and friend for an ITM Fishing Show programme on Doubtless Bay snapper from the rocks. *SAM MOSSMAN*

conditions, over a period of several months. As a consequence, the spawning season tends to start slowly, accelerate rapidly, and then ebb away as spawners and non-spawners mix together in the warmer summer waters. The peak usually lasts a few short weeks, when a large percentage of snapper are all spawning at the same time. Whether they are just starting to spawn or just finishing, the result is a very large aggregation of snapper in one small area at the same time.

This peak typically occurs in the last weeks of November and the early weeks of December. The spawning season of 2006–7 was different. Spring started out warm but quickly became cool and dry. Winds from the north gave way to winds from the west and south: cooling winds that blew warm surface waters offshore. Coastal water temperatures plummeted. This offshore El Niño weather pattern took hold and delayed the annual spawning season by many weeks.

Not until January did the snapper spawning run get fully underway. The sou'westerly stopped, and gave way to an extended period of northeasterlies. Here was the meteorological shift that everyone had been waiting for: summer was finally here. The problem was this meant it began at the start of the Christmas and New Year holiday season, bringing hundreds of recreational snapper fishers to Northland holiday destinations.

The result was one of the largest aggregations of recreational craft ever seen in Bay of Islands waters. The huge flotilla extended from Ninepin Rock to Whale Rock, and out to the reefs off Moturoa. Initially, the water stayed cold and not many people caught fish, but as the wind kicked in a little harder from the east a wall of warm blue water came with it and the snapper went into a frenzy. As if a starting gun had been fired, snapper invaded the Bay of Islands and came on the bite as one inconceivably massive school. Everywhere people were catching fish, many of them large. Sounder tracings recorded a massive layer of snapper that stretched over more than five square kilometres. This was snapper fishing as few had seen it before.

It was an event that gave us a visible understanding of the physical size of some of our northeast coast snapper fisheries. Snapper are, after all, something of an unseen fish. They rarely school at the surface and are not inclined to swirl around parked boats like inquisitive kingfish. Mostly they are shy and tend to retreat into waters of depth and low visibility, nor are they often seen en masse. In this situation, the sheer volume of widespread fishing success gave some idea of the massive numbers of fish involved. Literally hundreds of people were catching fish at once.

The Bay of Islands in January, with a burning sun and a warm wind coming off the ocean, was a portrait of prime snapper country: a multitude of sandy islands and shell-lined channels running between them. Beyond these lay vast areas of shallow seagrass beds and scallop flats, and further on an unseen array of mangrove creeks, secluded bays, safe anchorages and quiet waterways: an abundance of snapper nursery waters. Overall, a prime spot for adults to spawn and leave behind developing offspring.

The strength of the snapper runs in the Bay of Islands and further up the east coast at Doubtless Bay, Karikari Bay, and Great Exhibition Bay, is a reflection of the quality of the nearby nursery waters. Snapper don't make conscious

Before refrigeration, drying snapper in the sun and wind was the common way of preserving them for later use. Two men with a catch of Northland snapper in the 1920s.

ARTHUR JAMES NORTHWOOD. NORTHWOOD COLLECTION, ALEXANDER TURNBULL LIBRARY, WELLINGTON, NZ. G- 6326-1/1

decisions about where they're going to drop their eggs. Strong juvenile populations lead to strong adult populations and ongoing runs of big spawners. The nursery waters are a crucial factor in this whole sequence of spawning events.

THE RECENT PAST

If this was an example of a contemporary snapper spawning school, then the spawning schools of yesteryear must have been even more incredible. Stories of how large the fishery used to be are not hard to find: tales of large patches of water turning red as the snapper came to the surface to spawn can be a little hard to fathom, as are tales of big snapper schools swimming right into the upper Bay of Islands waters with school sharks and bronze-whaler sharks harassing them all the way. In fact commercial fishers used to drag for snapper off the upper harbour beaches, sometimes right in at Waitangi, and cursed the sharks that were an ever-present menace in their nets.

Why snapper do not penetrate as far into estuarine waters for spawning purposes in modern times is hard to ascertain. The principal spawning zone is now out around the islands. Either the bulk of those shallow-water spawners have been caught and taken out of the population, or some change in water quality may have occurred in the Bay of Islands in the last 60 years. Possibly noise is keeping the snapper schools away. The inshore channels of the Bay of Islands are now ocean highways, where the howl of outboard motors is incessant.

It could also be simply the depletion effect we see in most fish populations. The instant a fish population is reduced, a retraction of fish from the peripheries of the range is noted. Hence you get fewer snapper these days up mangrove creeks and over ocean beach shellfish beds in the middle of the day at any time of year. Muddy grey estuary holes used to produce big snapper in the days when vehicle transport was a luxury. Ray Doogue (author of *Hook, Line and Sinker*) reported catching numerous snapper in the middle of the day off sandy beaches in the 1950s. Both are considered rare events these days. But the retraction of fish from the limits of their range is very real. Present snapper fisheries probably run at 25% of virgin biomass.

For a while the snapper fisheries of the far north gave a glimpse of what the past may have been like. Even during the 1970s stories came out of the Parengarenga and Rangaunu Harbours that would make people shake their heads in disbelief. One favourite story described large schools of snapper feeding over shallow shellfish beds on moonlit nights, swimming vertically with their tails flapping in the air. It was a good story: hundreds of large red tails swirling through oily-calm water waving at the moon. It may still occur today, but with the snapper population reduced the numbers of fish over the pipi beds are going to be far smaller and the occurrence perhaps not quite so eye-catching. A few swirls in shallow water at night will probably be passed off as mullet. A new reality of what snapper are and where they inhabit takes over – the 'sliding baseline' in action.

Another feature of the old fisheries that stands out was the abundance of large snapper. The population pyramid for snapper was far more top-heavy, with a lot of older, big adult fish – probably the most common large predator in

The pure white silica sands and clear waters of Parengarenga Harbour in the Far North make it one of the New Zealand's most picturesque snapper habitats. *SAM MOSSMAN*

many northern marine habitats, most commonly on coastal rocky foreshores. These fish had the size and powerful jaws required for crunching up many of the hard-shelled invertebrates found in these reef areas. They provided a common enough resource along the extensive rugged foreshores of Northland's east coast for a whole angling culture to evolve around them. Battle-hardened rock fishers would target these snapper with heavy rods, big hand-lines and large baits. Fishing specifically for large 'kelpie' snapper was very much an entrenched Northland activity.

The late Peter Mellsop of Whangarei was one of these fishermen. He would hike into big fish ledges with a sack slung over his shoulder and a heavy surfcasting rod at his side to fish using big tough baits among jumbled areas of rock and weed. His bait was far larger than anything most people would typically choose: often a whole small octopus or a big squid. This was cast, unweighted, into visible gaps between the rocks and weed. The bait was not allowed to drift free but retrieved just fast enough to keep slack out of the line and prevent the bait from snagging on the kelp. When a snapper grabbed the bait Peter would drop the rod tip and feed slack line to the fish. He often caught snapper of 10 - 12 kg this way, often on bright and sunny, calm days.

It was a system for extracting big snapper out of the kelp beds in which they hid, a specialised technique designed for a specific population of big fish that were hungry and widespread. Competition for available food resources amongst resident snapper would have been intense. Many local gamefishing boats used to park up for the night close to broken reefy ground and cast big baits back at the washes. 'Twenty-pounders' were commonplace and expected. Nobody considered them trophy fish.

Stories exist of large resident snapper sitting under shag colonies waiting for fledgling chicks to fall from their nests, or large snapper preying on moulting crayfish in extremely shallow water.

Such is the aggression of snapper that there are stories of them attacking skip-baits meant for marlin when passing close to the island washes. Veteran Bay of Islands skipper Snooks Fuller recalls trolling whole kahawai skipbaits near the Cavalli Islands:

'After a storm, around the outer rocks, the snapper seemed to be ravenous. If the boat took a turn around a rock and the troll bait sank a bit, the snapper would attack it and sometimes became foul-hooked; not necessarily huge ones either.'

Recently a large snapper was reported chasing schools of yellow-eye mullet around in light surf on Long Beach, Russell. Retirees walking their dogs that morning watched from only metres away as the snapper charged back and forth through the small waves with its fin and tail in the air and big sprats scattering from under its nose. Many watching passed it off as an aberration: an unusual event not likely to be witnessed again in a hurry. However this sort of activity was probably once common. Yellow-eye mullet schooling in shallow water represent a large protein source for any population of big snapper. They probably attacked such baitfish regularly in the past, and may still do so today under the cover of darkness. The most unusual aspect of this event was that it occurred in bright morning sunlight, reflecting once again how our collective perceptions of what snapper are and how they should behave have changed. This change has occurred purely because of the modern-day limits of the fisheries. If snapper biomass was allowed to rebuild to even forty or fifty percent of what it was, it would create a fishery and fishing environment quite different to that which we see today.

Before humans, the largest snapper had few predators and only the limits of their food resources restricted the size of the adult population and its breeding success. Reef fish and other small fishes of the ocean floor would have been preyed on heavily and even small snapper would have faced their greatest threat from snapper larger than themselves.

THE NEW REALITY

The present snapper fisheries from Bream Tail through to North Cape are still surprisingly robust. Without getting into the global warming debate, it has to be said we have had some warm years in recent times which have undoubtedly benefited the snapper. A strong correlation has been shown between lingering Indian summers and healthy stocks of small snapper. Provided the total snapper catch stays as it is, the prospects for the fishery in the future is are relatively good.

In recent times, however, a notable shift in commercial fishing has occurred towards the far north, while the angling pressure in the more tourist-oriented Bay of Islands has become heavily recreational.

DARRYL TORCKLER

NINETY MILE BEACH AND THE NORTHWEST

06

Sparsely populated and free of large water-polluting industries and runoff from rivers, the wild and beautiful Ninety Mile Beach – Te Oneroa a Tohe – stands out as a mecca for surfcasters.

When the wind blows from the easterly quarter, Kon-Tiki and kite exponents fish the beach, and when sea conditions are favourable Ninety Mile Beach attracts many eager surfcasters anticipating a fine snapper for the table. They have a lot of spots to choose from; between the southernmost point of Ahipara and Scotts Point in the north lies 84 kilometres of hard-packed sand. The ever-changing holes and channels, carved by the sometimes fierce currents of the Tasman Sea, are sought-after fishing spots.

Washed constantly by plankton-rich waves, this beach hosts huge beds of tuatua and, to a lesser extent, toheroa. These shellfish grow plump and tasty, and attract many fish species to this beach, none more so than the snapper.

According to old Ninety Mile Surfcasting Club records, the average Ninety-Mile snapper has never been exceptional, mostly 1–3 kg 'schoolies'. The best fish recorded from any club field day was 9.2 kg, while an average winning fish would be 5–6 kg. There was one intriguing reference however:

'Two snapper were captured on the 90-mile in 1943 by Shirley Gardiner and Stan Harris. The largest weighed 39 lb scaled and gutted and was 4' 3' (1.30 m) long. The smaller fish weighed 30 lb. The larger fish was in poor condition but edible, and fed six families. The fish were not caught by line but stranded in a pool and caught by hand.'

With names, dates and measurements, this report has the ring of truth about it, even though the length given for the larger fish exceeds the longest recorded snapper (1.05 m) by a considerable amount, and is probably the overall length rather than fork length. Even so, the green weight of this fish was probably well over 40 lb. However, the comment that the fish were 'stranded in a pool' is not really consistent with the sand sections of the beach, being more typical of rocky territory so perhaps these fish, like most large snapper, were living in heavy foul.

At Ninety Mile Beach, a 9 kg fish is considered a monster, while on the east coast it takes a fish of 15 kg to earn that title. Given that west coast snapper tend to grow faster than their east coast counterparts, the reason could be habitat. Big snapper seem to prefer heavy foul, and the country a little to the north around Cape Maria van Diemen is renowned for producing very large fish. The open, surf-swept sands of the Ninety Mile are a much more suitable habitat for school snapper.

Ninety Mile snapper have to constantly swim against the swift currents that make this beach such good snapper habitat. It is therefore rare to catch a snapper off this beach that is not in prime fighting condition.

EARLY DAYS

Ninety Mile Beach was an access route and a major food source for the early human inhabitants of the far north. Later, the huge kauri forests and swamplands attracted large numbers of loggers and Dalmatian gumdiggers who also fished this beach for food.

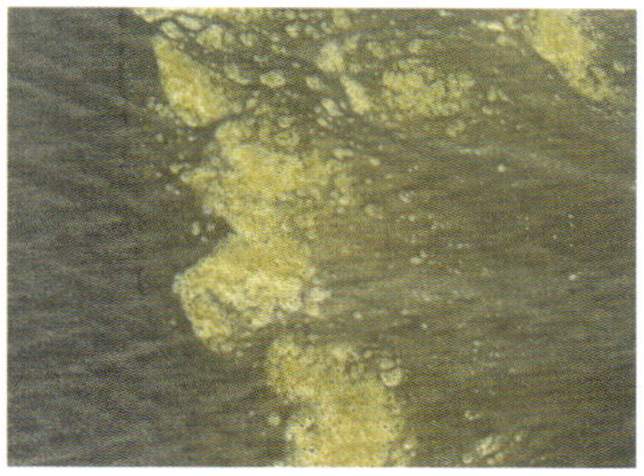

FAR LEFT: The late Charlie Norris of the Maori Battalion fought at Monte Cassino in World War 2. Here he enjoys success fishing Ninety Mile Beach on RSA Day, when local surfcasting club members take the old soldiers out fishing. *DAVE HARRISON*

LEFT: Where it all begins – the plankton that is the basis of the ecosystem on west coast beaches feeds shellfish like toheroa and tuatua, and fish including mullet. *SAM MOSSMAN*

CENTRE: Wetsuits are the approved beach wear for fishermen on 'the 90 Mile', both for warmth and safety. Manu Hepa with a fine catch from the surf. *BRUCE BASHER*

BELOW: Dave Bennett, winner of the Ninety Mile Beach Snapper Classic in 1991, with his catch – and a cheque for $50,000. *COURTESY OF TONY AND JEAN BRLJEVICH*

In times of hardship, snapper, toheroa and tuatua were a staple item on the dinner table. Alice Evans, in her 1981 book *Mount Camel Calling*, recalls: 'Back in the 1930s, though, it was our beach – wild, remote and lovely, a place to go for picnics, to gather the then-plentiful toheroas, to catch snapper, or just to be alone with nature.'

In the early 1900s, when kauri gum was a major export, large schools of snapper were often seen feeding in the surf. They would come into water so shallow that their tails would stick out as they fed. People looking for a feed of fresh fish would judge their size by how big their tails were. If they looked too small the fishermen would move on and keep searching until they found bigger ones. Casting only lightly-weighted tuatua or fish baits meant that often quite a few fish could be caught before the school was spooked.

While fishing techniques and equipment were not as sophisticated as todays, a good cord handline still worked well, as did set and drag nets. Rangoon cane rods and centrepin reels, while not good casters by today's standards, caught their fair share and became the surfcaster's weapon of choice.

FISHING WITH A BANG

The gumdiggers developed an even more effective method that required less effort. A lighted stick of dynamite was thrown into a school of fish, providing a good, easy feed. Nothing was wasted: if it couldn't be eaten fresh, it would be either dried or salted for later.

The government of the day outlawed the practice after an unfortunate accident. One man's method was to walk to the water's edge and light the dynamite off his cigar. One particular day, after a heavy session on the local home-made alcohol, he threw the cigar instead of the dynamite. The explosion blew off his arm and severely injured the rest of his body.

ON THE BEACH

Despite an ever-increasing human population and consequent increasing fishing pressure, 'tailing' schools of snapper were quite common well into the 1960s. However, increasing commercial pressure by bigger and more powerful trawlers towing larger nets slowly reduced numbers to the point where this sight is rarely if ever seen today.

Bob Guy, a husky, white-haired Awanui farmer now in his seventies, remembers wading among tailing snapper in knee-deep surf, such occurrences being 'not uncommon in the old days'. He last saw tailing snapper in the surf near Glinks Gully in the Opoto area in 1971 – 'the last year it was legal to harvest toheroa'.

Fewer fish now meant that surfcasters needed to be more cunning and put in a lot more effort. Needless to say, this extra cunning came in different forms. A classic example was the method used by well-known fisherman, the late Gil Henderson. Gil lived in Awanui and was regularly seen fishing on his favourite Ninety Mile Beach spot, where he amassed a huge amount of skill and knowledge while fishing there in the 1970s and 80s.

ABOVE LEFT: In 1970 the Ninety Mile Beach Fishing Contest began with a mass start and beach race. This was eventually stopped for safety reasons. *COURTESY OF BOB GUY, NINETY MILE BEACH SURFCASTING CLUB*

ABOVE: A five-day Ninety Mile Beach surfcasting contest was first run in 1982 by the Brljevich family (left to right) Ross, Tony, Jean and Paul, in 1981 as an event to promote the Far North District, and ran for 25 years. *COURTESY OF TONY AND JEAN BRLJEVICH*

LEFT: By 1973 the surfcasting contest was hugely popular, with vehicles and competitors as far as the eye could see. *NORTHERN ADVOCATE/BOB GUY, 90-MILE SURFCASTING CLUB*

FAR LEFT: The early Ninety Mile Beach surfcasting contests were run by the local surfcasting club. £600 was big money in those days – more than a year's pay for some. *SAM MOSSMAN/BOB GUY*

Gil liked to fish spots where the water was about a metre deep, on a rising tide, over a tuatua bed. He would take his caravan up the beach at low tide and have a sleep until the tide came in. His favourite method was to cast a fish bait on to the tuatua bed to catch snapper feeding in the incoming tide.

Others took to using neoprene wetsuits and wading out to fish off sandbars for an hour or two either side of low tide. Conditions on Ninety Mile are often heavy and can involve deep wading in up to two-metre swells. Seas of 1–1.5 m are considered to be quite fishable. Serious Ninety Mile surfcasters think little of wading up to chest deep, and are sometimes totally submerged by swells. These days wetsuits and lifejackets or inflatable flotation devices are used for safety and comfort, but before wetsuits early surfcasters used a 'port wine and Old Parr whisky' wetsuit, according to Ninety-Mile veteran Bob Guy.

A good watch had to be kept for stingrays and eagle rays sunning themselves in the warm shallow water, but deep wading allowed anglers to cast out much further. Kaitaia accountant Dave Harrison is one of the few to have been injured by a stingray when wading there. He stood on a ray which thrust its barb right through the calf muscle of his right leg and lacerated his left foot below the ankle. Dave suffered much blood loss and has a deep 80-cm scar to show for his encounter.

Generally though, deep wading is well worth the effort. To keep fish secure while wading back to shore, some have large pockets built into their jackets keep fish in. Wading is still the most often practiced technique today.

Tackle has come a long way. The Rangoon cane rod has yielded to rods made from composites of modern materials such as carbon fibre, boron and Kevlar. Reels now feature long-cast technology and are far more robust. The early centrepin type has evolved into the side-cast reel. The Alvey brand is perhaps the best known of this type, and its robust simplicity stands up well to the exposure to salt and sand.

Time of year and day has become a lot more important for surfcasters now. The early morning and evening change-of-light periods are best. Because of swift rips and deep holes, wading during the darker hours must be done with extreme caution. The larger snapper are also more prolific after the water warms up in early October, through to about June, with the prime time from March to May.

In the Kiwi tradition, the far north is inundated by holidaymakers during Christmas and New Year. Very few head home again without sampling the wide array of seafood the area has to offer. Many locals of the district visit the Ninety Mile for their break, and count a fillet or whole snapper sizzling on the barbecue as a highlight.

CLUBBING TOGETHER

The Ninety Mile Beach Rod and Line Club was formed at Ahipara on 3 March 1954. The name was changed to the Ninety Mile Beach Surfcasting Club in 1956, and this club has been synonymous with surfcasting for snapper there.

The physical demands of fishing the Ninety Mile preclude a large junior membership, and likewise there are only a few older members – the big surf is too tough. The core of the club is around 50 working-age adults, with a fringe membership that comes and goes. The club holds regular field days on the beach during the snapper season.

In the late 1960s and early 1970s, the club ran a snapper surfcasting contest on the beach, which has now become a huge five-day tournament. The club-run contests were big too, and initially opened with what amounted to a mass car race up the beach to get to the best spots. This part of the event was eventually stopped when it began to get too dangerous.

THE NINETY MILE CONTEST

The event became too much for the small fishing club, but the value of the event to the far north had been clearly established, and the torch was passed. Now, in February or March each year, Ninety Mile plays host to what is arguably the world's premier surfcasting contest. Held over five days, this event attracts a thousand hopeful surfcasters from both New Zealand and overseas.

The five-day format was first conceived by Tony and Jean Brljevich in 1981 as an event to promote the far north district. It first took place in 1982 when the first prize, for the heaviest snapper was $15,000. Since then the prize has soared to $50,000 with a total prize pool of $250,000. While smaller fish have won, those taking out the top prize have mostly been prime specimens of 6 to 9 kg.

Now known as the Lion Red Snapper Classic, this contest has become a New Zealand icon. At the 2006 event the Brljevich family stood down from their organisational duties and handed the reins over to Srecko and Yvette Urlich. A lot of visiting Australian anglers are amazed that Kiwi surfcasters are able to catch snapper off a beach. When the Australian Alvey team visited a few years ago, Bruce Alvey declared, 'Back home we have to go out in a boat to catch a snapper – you guys have got it made!'

Little wonder that surfcasters from all over the country return time and time again to this awesome place.

WEST COAST HARBOURS

The long sandy stretches of this coast are interspersed with harbours, mostly guarded by fierce currents, changeable channels and bars that have become notorious as the graveyards of ships and sailors – the Hokianga, the Kaipara, and the Manukau have all seen a long list of nautical disasters, most especially in the sailing days, when ponderous square-riggers were caught by the prevailing westerly winds on an unforgiving lee shore.

What may have been a bane for sailors has created a productive fishing ground for anglers. When the weather and sea conditions permit, smaller boats can go out over the bar, often experiencing wonderful fishing out on the sand, or on isolated patches of foul, the positions of which are often a closely-guarded secret.

The harbour entrances can also provide excellent snapper fishing at the right time, with some of the most famous

TOP: Trailerboat fishermen launch off the beach near Dargaville, Piha, and other spots along the northwest coast. You really have to know what you are doing though…
STEVE WATERS

ABOVE: The fertile mudflats of the Kaipara Harbour are a prime feeding ground for fish, including snapper, and the extensive channel system provides many a snug mooring for boats that ply these waters.
SAM MOSSMAN

spots having fearsome names like Destruction Gully on the Manukau and The Graveyard on the Kaipara. This latter is a popular spot that can be reached from various parts of the vast, convoluted Kaipara Harbour, including Helensville on the southern extremity, where a number of charter boats are based. A trip out to the Kaipara Heads from Helensville can take over three hours on some boats, but the ride and the camaraderie of regular anglers are part of the experience.

There are some fierce rips where this huge harbour funnels out through the heads. Charter fishing groups tend to use substantial tackle (15 kg is popular) and sinkers of 16 to 20 oz to hold bottom in the fast water. Well-organised groups minimise tangles by all using lines of the same thickness and sinkers of the same weight. Here ultra-thin 'superbraid' lines have their advantages although they are hard to untangle when the inevitable does happen. Still, nylon and 16-oz sinkers are a big step forward from the old days when heavy handlines were weighted with cast-iron window sash weights that weighed some pounds. Snapper in this area tend to be of 'pannie' size, but there are usually plenty of them and everyone goes home with a feed.

In the summer months, school snapper move in to the rapidly-warming west coast harbours to take advantage of the rich feeding grounds in the channels, banks and muddy mangrove creeks. They are seldom very big fish, but these 'schoolies' are ideal table fish, and the sheltered harbour waters are well suited to small boats.

The tradition of shallow-water snapper fishing in west coast harbours is an old one. The late Temple Sutherland described an episode that probably took place in the 1930s, in his 1963 book *Maui and Me*:

'After we had gone about half a mile Doug told me to throw the pick out. The water covered the top fluke with about two inches to spare.

'"Bit shallow," grunted Doug. "We'll let her drift over the channel."

A few minutes later, using an oar as a dipstick, he confirmed that we had the big end of a fathom [two metres] of muddy water under us.

'"That's better," he said, hauling out the makings. "Can't beat a bit of depth for this game. It's a good channel this."

He rolled, and lit his smoke; impaled two hunks of lightly-smoked eel on his hooks, and tossed them over the wall.

A touching act of faith, I thought. Our distance from the shore was illusory. Even from this 'deep' spot we could just about walk home without getting our collars wet. Another factor was the temperature of the water. After inching in over hundreds of yards of sun-baked mudflat it wasn't just warm, it was hot. No fish was going to get itself parboiled in hope of finding a piece of smoked eel.

But, incredibly, Doug's finger was already twitching.

'"Better make a start," he advised. "They're beginning to come about."

And the next moment he'd made good his prediction by boating a medium-sized snapper that felt quite warm to the touch. A couple of minutes later I was busy with one myself, a five-pounder putting everything into the short, sharp struggle. Nothing wrong with this for sport, I had to concede.'

Fishing off the northwest coast by boat is limited by the exposed nature of the area. Most of New Zealand's weather systems come from the west, which makes for a high-energy coast with big surf beaches and dangerous, high-current bars on what reasonable-sized harbours there are.

There is also the option of beach-launching if conditions permit, and any prolonged period of easterly-quarter winds can see boats braving the surf at access points from Scotts Point at the top of Ninety Mile Beach, all the way south to Muriwai Beach on the back door of Auckland.

However, the difficulty of access and the logistics of boat fishing this stretch of coast afford its snapper some protection, and anglers who do get out (particularly during the spring and summer) can find some exceptional fishing. Spots are usually described in terms of water-depth (e.g. 'the fish are at 55 m off Piha' means they are in that depth of water, not 55 m off the shore) as much of the fishing is over open sedimentary bottom where the fish move about in schools. Areas of foul are relatively few along this coast, and anglers that know of unmarked reefs tend to keep quiet about them.

The West Coast just north of Port Waikato on a calm day. Often, however, heavy weather on this coast is a danger to both boats and shore fishers. *SAM MOSSMAN*

DARRYL TORCKLER

HAURAKI GULF AND FIRTH OF THAMES

07

The Hauraki Gulf and Firth of Thames region may be close to a perfect habitat for the flexible and versatile snapper.

It seems that the biggest populations of snapper live in large embayments with sedimentary bottoms, good and varied food supplies, protected nursery areas, and most importantly, thermally-stable water masses that can reach and hold the key water temperatures of 18°C or more that are required for a successful spawning. The Hauraki Gulf can sustain these conditions for up to four months between December and April, enabling serial spawning – individual fish spawning several times over a period of months, rather than in just one big hit. This enhances genetic diversity, because this way they have more than just one mate, and increases the chance that at least some of their eggs will strike optimum conditions.

Nursery areas for young fish are plentiful in the Gulf and the Firth, and in addition to the extended spawning season, the snapper here have the luxury of late maturity, so when they first breed they are about twice the age of snapper from places where ideal spawning conditions are much less frequent. The diverse habitat also provides a great range of feeding opportunities for a broad-niche feeder like the snapper, ranging from the shallows of the upper Waitemata to the deep-water reefs beyond Great Barrier Island.

Spawning success and diverse feeding opportunities make Gulf snapper one of the most robust stocks. Evidence from middens shows they were important to early Maori in the region. Snapper have been fished hard (often overfished) commercially for more than a century, and in recent times have also been under pressure from the nation's largest concentration of recreational anglers. Despite all this, when fisheries management regimes give them half a chance, snapper stocks in the Gulf/Firth area keep bouncing back.

THE BLUE BOATS

Recreational demands in the late 1950s and early 1960s saw the development of a charter fishing fleet that for a 'quid' or so each (£1, equivalent to $2 when decimal currency was introduced), would take groups of amateur fishers out for a half-day or day of snapper fishing. Groups of 40 to 90 fishers would board party boats such as the *Florence Kennedy*, *La Reta* and *Shenandoah* from Princes Wharf at the bottom of Queen Street, and often return with big catches of snapper proudly displayed along the rails of the boat. Later, many of these launches were painted blue and the fleet became known as the Blue Boats.

For their quid, anglers got a section of the rail to fish from. A sack was tied to the rail to put their catch in, and handline and bait were provided. Temple Sutherland, in his 1963 book *Maui and Me*, recalls:

'...the skipper got on his marks and cut his engine. I asked him what the bottom was like.

'"Clean, with beds of horse mussels," he said. "We ought to drop onto a few here."

'Even as he spoke the first of the predicted few were coming aboard and being lowered in to the sacks slung on the

ABOVE: Auckland Harbour as painted by Arthur Edward Williams in August 1864. Local Maori unload a catch of snapper from their waka. European ships are anchored in the roadstead, with North Head behind.
ALEXANDER TURNBULL LIBRARY, WELLINGTON, NZ. B-045-012/013

LEFT: With a wealth of sheltered estuaries, islands, reefs and channels, the Hauraki Gulf is ideal snapper habitat.
SAM MOSSMAN

rails. Excited voices rising to high-pitched squeals of delight as a party of teenagers boated a kahawai. The 'regular' in the panama broke his duck with a six-pound snapper, and sacked it without emotion.

'Finding the line I'd brought a trifle short I accepted one offered to me by the skipper, a strong nylon, rigged with a running sinker, long leads, and looped hooks for easy detachability. And within ten seconds of my sinker's touching the bottom I was fetching up my first snapper. Seeing I was likely to be busy for a while, the skipper resumed his book. His job was done for the moment. He was over fish. Everyone was happy. The deckhand, having served a cup of tea, was on his rounds, giving a hand or advice where needed. Round us on the calm water sooty shearwaters skirmished and bickered over bits, diving deep into the clear blue on the trail of the disappearing baits, and staging tugs of war with the pieces of skin that came their way.

'With five fair-sized snapper in my sack I took time off for lunch, well on my way to cracking even. *Shenandoah* lay almost without motion, the light breeze scarcely ruffling the water. The one cloud in the sky poised like a permanent halo over Little Barrier to the northward. To the west, Tiritiri Island stood out from the heat-hazy coastline. The teenage party was busy with lunch and tubes of sunburn lotion. The man in the old panama fished on steadily, putting an ever-increasing bulge in his sack. He had cracked even long ago and was on the way to reaping a nice little profit for his day.'

On a good day many would return with a catch of fresh fish worth more than their charter fee, essentially having a good day's sport for free. There was also the opportunity to back yourself in a sweepstake for the biggest fish. In those days a little dockside commerce was not unheard of, either. Sutherland continues:

'I asked the skipper if there were any restrictions on selling surplus fish.

'"Plenty of them," he said. "It's illegal to sell fish caught on the boat without a commercial fishing licence. It's illegal to hawk fish without a licence; to sell in public places and so on. My only responsibility is to see there's none sold on the boat or at the boat side."

'I gathered that quite a lot of fish did in fact change hands. On Sunday afternoons when all the boats berthed about the same time there was often a huge crowd to see them come in, sightseers and people hoping for a bundle of fish at a bargain price. When you noticed a little scrum going down here and there through the crowd it was usually a sign that willing buyer and willing seller were transacting a deal.'

In more recent times most of the slow displacement-hull charter boats have been replaced with more speedy craft which depart from many places around the Hauraki Gulf. The handlines and sacks have been replaced with rods and iceboxes, but the hopeful faces, the islands of the Gulf and the snapper remain the same.

SNAPPER AND THE CUP

Closing the inner channels of the Hauraki Gulf to commercial fishing in summer has seen a great improvement in the

LEFT: Starting in the late 1950s recreational demands resulted in the emergence of a charter fishing fleet (such as the *MV Florence Kennedy*, seen here) that would take groups of amateur fishers out for a half day or day of snapper fishing in the Hauraki Gulf, often returning with big catches tied along the rails of the boat for public display.
ARCHIVES NEW ZEALAND/TE RUA MAHARA O TE KAWANATANGA WELLINGTON OFFICE AAQT 6401, A75374

BOTTOM LEFT: What every angler seeks. Joe Jackson with a trophy snapper from the Mokohinau Islands. *SAM MOSSMAN*

BELOW CENTER: When the mussel barges are harvesting their crop, the dropper lines are pulled through a stripper. The unwanted portion, including tasty snapper treats like crabs and small crushed mussels, is discharged over the side, setting up a feeding frenzy for snapper that recreational anglers, with care, can exploit. *GRANT DIXON*

BOTTOM RIGHT: Bird activity is keenly watched for in the Hauraki Gulf, as there are often plentiful snapper feeding under such melees. *SAM MOSSMAN*

OPPOSITE, CLOCKWISE FROM TOP:
The magnificent Mokohinau Islands, the 'Mokes' to locals, mark the outer reaches of the Hauraki Gulf, and a hot-spot for big snapper. *SAM MOSSMAN*

Soft plastic lures are just as successful on 'pannie' snapper in the hard-fished Rangitoto Channel as at offshore hot-spots. *SAM MOSSMAN*

Fishing for snapper with soft plastic lures has become very popular in recent years, and it is little wonder with results like this Little Barrier fish caught by Dean Henderson. *SAM MOSSMAN*

Heli-fishing involves using a helicopter to access offshore islands or hard-to-get-to headlands, from where the anglers rock-fish. An ideal way to get quickly, and with little physical effort, into isolated spots which can hold big snapper – if you have the dollars. Gary Larkin, Sarah Thompson and Peter Elliot with a few keepers. *SAM MOSSMAN*

numbers of snapper available, especially to small-boat owners whose craft have limited range.

Another totally unintended benefit came during the two years when the America's Cup regatta was held in the Hauraki Gulf. Coincidentally, the course was set over an area fishermen call the 'worm beds', an important schooling, feeding and spawning ground. For an entire season fishermen were unable to work the area for much of the time, giving the snapper a good rest. Although most of the spectator fleet were not permitted to anchor or fish, sounders showed very heavy fish concentrations in the water under the course.

MUSSELING IN

Another man-made opportunity is provided by mussel farms. Snapper have cashed in so quickly on these well-stocked new 'shellfish supermarkets' that most mussel farmers welcome anglers who help thin out their numbers, especially in the Firth of Thames and the Marlborough Sounds.

Fishing in mussel farms requires care and careful navigation and anchoring (to avoid tangling or damaging mussel lines) but the action can be full-on, especially when the mussel barges are harvesting. Besides the mussels, there is a huge amount of marine growth and other organisms attached to the dropper lines. The ropes are pulled through a stripper, and what is peeled off gets put through a tumble-cleaning process. As a result all kinds of tasty snapper treats like crabs and small crushed mussels are thrown over the side, creating a massive berley trail and setting up a feeding frenzy. Sometimes the snapper are so thick around the outfall that their backs show above the surface and the water, stained brown by the wash-out, takes on a red tinge from the sheer density of fish. It also helps that these snapper are already accustomed to engine noise. Careful fishing in this zone can see plenty of action but it's important to check the etiquette first with the barge operators.

Some mussel farmers offer charter trips that may combine fishing with a lesson in the workings of the mussel industry, sightseeing, and a barbecue. Mussel farms are not 'trophy' fisheries, most snapper being small to medium-sized panfish. This is perhaps an advantage however, as the many snags in a mussel farm make handling a big fish difficult.

PLAYING WITH YOUR FOOD

It should be no surprise that the Hauraki Gulf region is a hotbed of developments in snapper sportfishing, with the region's huge population and popularity and abundance of snapper. A case in point is fishing with hard metal jigs. There are references to catching snapper on metal jigs dating back to about 1956 in books by the 'godfather' of New Zealand sportfishing, Ray Doogue. However, the development and fine-tuning of the technique owes much to early experimentation by people like Bruce Riley, Frank Perry and Eric Morman in the late 1980s, in the Hauraki Gulf.

Essentially, the technique consists of dropping a fish-shaped metal jig to the bottom from a drifting boat, and jigging it up and down with the rod. Although it has since rather fallen out of fashion, this technique is very successful

on bottom and midwater-schooling fish, around structures, in channels and under bait schools marked by feeding birds, and has been used to catch many fish species all over the country.

The same may be said of fishing with soft plastic lures for snapper, which has recently become extremely popular. Soft plastic lures have been fished on and off by some anglers over the years, but the method did not find widespread acceptance till after a range of tackle developments including scented biodegradable soft plastic lures and fine-diameter, low-stretch polyethylene lines. And again, perhaps by sheer number of anglers and the fact that the key tackle importers are Auckland-based, the Hauraki Gulf is where a lot of the development has been taking place.

Soft plastic lures first found favour in American freshwater fisheries, and from there the technique spread to salt water. In Australia, the soft plastic revolution was largely based around inshore saltwater fish such as bream and flathead, and even saw the establishment of a big-money professional tournament-fishing circuit, similar to the American bass-fishing circuit.

In New Zealand conditions are a bit different: fishing is often deeper and there is more current. Local anglers have adapted their own techniques. Heavier leads, the worm hook and running sinker, and various ledger-style rigs have extended the application of these initially shallow-water lures.

Snapper have proved an ideal target for soft plastic lures. Many anglers are only now coming to realise just how aggressive these fish can be towards baits of this kind. During the winter months, fish that hold up in heavy, shallow foul fall victim to soft plastics cast among the rock and weeds, and when they move out into the open and school up, drift-fishing and working lures along the bottom is very effective. When schools of baitfish are being worked by birds and dolphins, large groups of snapper are often underneath. The fishing can be phenomenal in this situation as large fish in midwater intercept the sinking lures.

Soft plastics have proved to be an addictive and versatile way to fish. The light tackle and fine-braid lines used to get the best from these lures mean anglers also get great sport and excitement out of catching fish that would have passed without comment on traditional, heavier gear.

THE FURUNO

With a prime snapper fishery and huge number of anglers from the Auckland urban region, it is no wonder that the Gulf saw the development of a huge and very successful fishing contest, the main target of which was snapper. The name changed a little over the years depending on the identity of the co-naming-rights sponsor, but to fishermen far and wide it was known as The Furuno, the front-line organisers being Mike and Marrisse Hodson and Keith Patterson of Electronic Navigation Ltd, the Furuno agents.

The event was first run in 1986 at Waiheke Island, shifted to Motutapu briefly, and then to Motuihe where it was based for many years before moving to Pah Farm on Kawau Island. Here, permanent facilities could be built and the

The Furuno fishing contest was run for 19 years from various venues in the islands of the Hauraki Gulf, and at its peak attracted over 3000 anglers. This shot shows the final venue at Pah Farm, Kawau Island.

COURTESY OF KEITH PATTERSON AND ELECTRONIC NAVIGATION LIMITED

Furuno was based for the remainder of its 19-year life. In its final year the prize pool exceeded $170,000 and at its peak the contest attracted over 3000 anglers.

The Furuno became a de facto snapper festival and included the prestigious Snapper World Cup event. The last Furuno contest was held in 2005 after a developer bought out the Pah Farm facility, but other large contests have since been established to fill the void.

THE CHEATS

With increasing numbers of contests competing to attract a static pool of angling entrants, prizes of cash and goods have reached high levels. Inevitably, rich rewards have lead to allegations of cheating: dark rumours of pre-caught snapper frozen then thawed for weigh-ins; fish kept captive in underwater cages until contest day; fish 'stuffed' to increase their weight after capture, and XOS specimens caught commercially then weighed in amateur contests.

Organisers usually go to great lengths to o eliminate cheating. At the Ninety Mile Beach Surfcasting Contest, anglers must wait at the water's edge for one of the patrolling marshals to check and mark the often still-flapping fish. On one occasion the organisers of the Furuno contest imported a torrymeter, a device to measure how long a fish has been dead.

A good weighmaster will spot the signs of a cheat. During a single hour's observation of the weigh-in procedure at one Furuno contest, a writer observed two anglers pulled aside and questioned. One had a bin of large trevally, but none had hook holes in the lips and several showed net marks. He eventually confessed to having caught them in a set net. Another angler admitted that not only had his very large john dory been caught commercially but he had also stuffed it with a small snapper to further increase its weight. Both these cheats were quietly dealt with, and few of the legitimate contestants would have been aware of what happened.

In 2001 organisers of another contest disqualified a large snapper weighed in for a $20,000 prize, the near-rotten fish having obviously been dead for much longer than the length of the contest, while a mullet and a kahawai head found in the its stomach were fresh. The cheats were prosecuted for false pretences, but the judge thought the scientific evidence inconclusive and dismissed the case.

BY SKY AND BY SEA

A new method of fishing for snapper came with the increasing numbers of corporate white-collar anglers based in Auckland: using a helicopter to access offshore islands or hard-to-get-to headlands, from where the anglers rock-fish, usually casting big unweighted or lightly-weighted baits around the foul.

Heli-fishing is an ideal way to get to isolated, little-fished spots quickly and with little physical effort – if you have the money. One Dunedin businessman caught the early flight to Auckland, helicoptered out to Great Barrier Island, fished all day, and then flew home that evening with a bin of fresh snapper fillets. A fabulous day with all the thrills of a chopper ride over some beautiful country, although an expensive way to get a feed.

First and foremost though, the Hauraki Gulf is a boatie's paradise. A prestigious overseas cruising magazine once rated the Hauraki Gulf along with the Bahamas and the Greek Islands as one of the world's top three cruising grounds. For owners of kayaks and dinghies, or multi-million-dollar yachts and launches, there is water to suit, from sheltered harbours and coastlines, right out to remote islands rising from the blue water of the outer Gulf. On a calm summer's day, particularly on a weekend, many thousands of boats of all sizes and types are out there, and many of them are going fishing. Their intentions vary, from highly-organised missions focused solely on catching fish, to simply dangling a line at the day's end in the hope of a bonus fish for dinner. But they all have one thing in common: everyone wants to catch snapper. And in this snapper Valhalla, most of them do.

JULIE-ANNE CUNNINGHAM

THE BAY OF PLENTY

08

The Bay of Plenty, like many places in New Zealand, was named by Captain Cook as he sailed across it during his first voyage in 1769. Local Maori, of course, already had their own names for these places.

Cook did not land but observed that it was well populated, looked fertile and its waters were full of fish, so he called it the Bay of Plenty. He also named, among other places in the region, Mayor Island (Tuhua to local Maori), the Court of Aldermen (today just known as the Aldermen Islands) and White Island.

Of them all, the Bay of Plenty may have been the most fitting name, especially regarding the fish. Even as late as the 1960s snapper were there in big numbers. In his 1967 book *Hook Line and Sinker,* Ray Doogue records encountering a massive school of surface-feeding snapper in Bay of Plenty waters.

'The skipper called out excitedly, "There are snapper everywhere – the whole school is snapper!"

And what a school it was, at least 200 acres in extent! There was no way of telling how deep the schooling fish went as every bait was seized before it had sunk 10 feet; I have not seen more fish or a run of better fish come over the side in such a short time. Within two hours we had boated about 80 snapper, which would have averaged 12 lb; the biggest was 21 lb and the smallest 7 lb.'

Those days are long gone now. The commercial excesses of the late 1960s and through the 1970s have seen to that.

A HISTORY OF BIG FISH

On the western side the Bay of Plenty is bounded by the mineral-rich, eroded volcanic landscape of the Coromandel Peninsula, with its scattered groups of islands – Cuvier, the Mercury Islands, Ohena Island, the Aldermen, Castle Rock, Shoe, Slipper, and others. Wide of Tauranga are Mayor, Karewa, Motiti, Schooner, Plate and Whale Islands, the Rurima Rocks, and a great many other islands and reefs. Off Whakatane is the offshore volcano, White Island.

Many of these places are famous for their big snapper. Mayor Island is renowned for its outsized reds; the world all-tackle record snapper was caught at Motiti Island; and Te Kaha Reef has produced many a monster.

Added to the massive amount of varied underwater geography are large, stable, sediment-based mid-depths ideal for spawning, temperatures ideal for good spawning survival, and many large rivers and estuaries that pump nutrients into the Bay, feeding the plankton which is the basis of all marine life.

The Bay of Plenty has a long history of producing outsized snapper. In 1993, Mark Hemingway's all-tackle record snapper was caught near Motiti Island off Tauranga. Mark was fishing with his father Kevin from the now-traditional inshore snapper boat, a 14-foot (4.5 m) 'tinny' (aluminium dinghy). Fishing an unweighted 10 kg line in just 6.7 m of water, Mark cast out a flying-fish head bait on a 7/0 hook. After the hook-up and initial run, the fish fought on the surface, and when it was brought near the boat its back was out of the water. The huge snapper tipped the scales at 17.2 kg (37 lb 14 oz) and was later accepted as the all-tackle world record for the species by the International Game

FAR LEFT: The volcanic heritage of the Bay of Plenty has spawned a huge variety of rocks, reefs and offshore islands that play host to some huge snapper. *SAM MOSSMAN*

LEFT: Dawn Irvine's 16.35 kg (36 lb) world record caught in April of 1998 near Slipper Island is, to date, the heaviest officially weighed snapper caught by a woman on regulation tackle. *COURTESY OF THE IGFA*

BOTTOM LEFT: In 1993, Mark Hemingway's 17.2 kg (37 lb 14 oz) all-tackle world record snapper was caught near Motiti Island off Tauranga. *COURTESY OF MARK HEMMINGWAY*

BOTTOM RIGHT: The big fish are often tucked up in heavy foul in shallow water. Carl Scott and young son Connor with a couple of beauties bound for Carl's smoker. *COURTESY OF CARL SCOT*

Fish Association. Hemingway donated his fish to science and it was aged at 45 or 46 years, considered young for such a large specimen.

The heaviest snapper ever officially caught by a woman on regulation tackle also came from the Bay of Plenty. In April 1998 Dawn Irvine, of Hamilton, landed the monster at the northern end of Slipper Island while she was 'just out to catch a feed' with friends aboard Henry Ralfe's *Ocean Lady* during a weekend visit to Whangamata. They were fishing in the middle of the day, without berley, in 30 m of water. Dawn was using standard snapper gear – 15 kg nylon line and a ledger rig baited with a modest piece of skipjack for bait. The fish was so strong she initially thought it was a kingfish. It weighed in at 16.35 kg (36 lb), a new world record that still stands.

WHITE ISLAND MONSTERS

Snapper are not common at White Island, perhaps because of the deep water that separates the island from the mainland, and perhaps because of the lack of suitable spawning water. There always have been some snapper there however, big old loners that find their way to this offshore outpost.

Whakatane charter skipper Rick Pollock has several stories to tell of giant White Island snapper. When he first started fishing there the main targets were the giant kingfish of the area. These fish were 'inexperienced' and the technique was first to get bait by trolling up small albacore and skipjack during the trip out from Whakatane. These provided just three baits each: two fillets and the frame.

Impaled on a 10/0 game hook, wire trace (some anglers used dog chain as a trace!) and 60 kg tackle, these crude offerings were pounced on by the unsophisticated monster kingfish of the day. On one of the early trips to Laisons Reef (between White Island and the Volkner Rocks) a whole albacore carcase was taken by a huge snapper, which on the official scales at Whakatane weighed 44 lb. This fish was caught before snapper were recognised as 'official sport fish'.

Ten years later, in 1989, Rick's charter boat *Pursuit* was at anchor for the night between White Island and the nearby Club Rocks. All the charter group had done their bait-catching duty and retired to their bunks – except one, a keen handline fisherman. At around 2 am Rick was awakened from his usual berth under the galley table by a faint cry for help. Investigating, the skipper found the fisherman battling a powerful fish that was wrenching his arms savagely as he hung grimly to his heavy handline.

'I thought it was a decent kingfish,' said Rick, 'but then I shone the torch over the side, and there was a huge snapper. It was over a metre long and around 35 or 40 lb. It made one last dive under the hull, the angler hung on – and the hook pulled!'

These days snapper are a little more common at While Island as a result of visiting charter boats bringing out small snapper from inshore waters in their livebait tanks and releasing them at the island. In one session during 2007 four were caught while fishing for tarakihi. Snapper which are smaller than any fish transported to the island have

LEFT: Spearfishing is a very selective way of harvesting snapper, as a fish, such as this prime table specimen, may be assessed before it is taken. *DARREN SHIELDS*

BELOW: Spearfishing was still a novelty in 1951, and equipment was basic. *EVENING POST COLLECTION, ALEXANDER TURNBULL LIBRARY, WELLINGTON, 114/395/05-G*

BOTTOM LEFT: The use of groundbait to attract snapper to an area is a successful tactic for spearfishermen. *DARREN SHIELDS*

BOTTOM RIGHT: Weedy guts are preferred habitat for big snapper, such as this trophy specimen. *DARREN SHIELDS*

been accidentally captured while fishing for jack mackerel bait, suggesting that some spawning does take place around this active volcano.

SPEARING SNAPPER

English-style hunting, on horseback, has been famously described as 'the unspeakable in pursuit of the inedible'. A maritime New Zealand version, ranking as one of the most unusual methods of catching a snapper, is documented by Ray Doogue in his 1976 book *Saltwater Fisherman*, from the upper Tauranga Harbour. However, in this case the rider was not unspeakable, nor the quarry inedible.

According to Doogue, in earlier days local Maori from Matakana Island used this method of catching the snapper that would wait out the bottom of the tide in shallow reaches of the upper harbour rather than expend their energy swimming back and forth with the tidal waters.

Riding a horse into the shallows, the hunter would spot a snapper and pursue it relentlessly. Eventually the fish would tire and duck into the turbid water stirred up by the horse, to hide. At this point, the rider would dismount and creep up on the fish, wielding a pitchfork. As soon as the snapper tried to move off from the clearing water it would be speared. The horse got some exercise, a wash and a swim, the rider got dinner, and the only loser was the snapper.

Commercial netting is reported to have cleaned out these shallow-water fish in subsequent years, and since World War 2, spearing the wily snapper has developed into quite a sophisticated business.

Spearfishing is a very selective method: the diver can see and select exactly the fish they want to take, and there is no collateral damage. Spearfishers say the big old snapper are extremely wary, an observation that matches the experiences of rod and reel anglers. This is described by Darren Shields, one of the best in the business of underwater hunting:

'Breathe slowly, fill your lungs, spit out your snorkel, and force your body forward into a dive. No kicking until your fins are below the surface; sink towards the edge of the weed covered rock, stop on the edge and peer slowly over. A big tail, the body, then those magnificent sweeping pectoral fins and, yes, a big humpy head – this is the ultimate sight for a spearfisher hunting snapper. Sometimes you can dive for days on end, visualising this sight, before finally seeing it for real.'

Spearfishing started back in the late 1940s when Leo Ducker, Eric Fryett and a few others pioneered the sport wearing woollen jumpers and thrusting spears with arm power before making catapult-type spearguns not too dissimilar to those used today.

They also were the first to use SCUBA for spearfishing, a practice not much seen nowadays, as you really need to be silent and unseen to get close enough to spear a snapper. It works fine for other fish, but is generally regarded as unsporting.

Monster or mock-up? Some fishy forensic work indicated that this snapper, while a very big fish, was not all it was claimed to be. *COURTESY OF NZ FISHING NEWS*

MONSTER OR MOCK-UP?

On the cover of the September 1985 issue of *NZ Fishing News* was a picture of what appeared to be a truly massive snapper, allegedly caught in the Bay of Plenty by a commercial fishing boat. The fish's weight was estimated as 'in excess of 60 lbs'.

The photo caused a sensation at the time, but was the fish as big as claimed, or was it as some have suggested: another example of the perspective trick where the fish is held well out in front of the angler? Certainly the commercial fisherman in the picture is standing behind the fish – but how far?

The following month a second photo of the fish was published with the fish held crossways in front of the fisherman, but with its tail and nose out of frame. The one clue is in the fisherman's fingers, which can be seen where he is holding the belly of the fish. In this photo, the width of the fingers at the first joint can be measured, and assuming the fisherman had an average width hand (90mm), the depth of the fish at the head can be scaled. This measurement can then be compared to the first full-length picture of the fish.

This method estimates the big fish to be 1.11 m from the snout to the fork of the tail. If this estimate is correct, it beats the largest reliably recorded fish of 1.05 m and 17.2 kg (37.9 lb).

There has been mention in literature of Australian snapper measuring 1.20 m and weighing 20 kg (44 lb), both of which are round enough figures to raise suspicion. However, using the length/weight ratios of these two last-mentioned fish to extend an existing snapper weight-length graph gives an estimation of the weight of the 'monster' fish in the photograph at about 18.7 kg (41.2 lb). Certainly a very big snapper, but a long way from the original estimate of 27.2 kg (60 lbs).

A group of big snapper hover over a rocky ridge where plankton-laden currents attract a myriad of fish life.
DANE HAWKER

In the 1950s many underwater clubs emerged. Spearfishing was very popular in New Zealand until a decline in t he 1970s, probably due to bad press about its environmental effects, although today its popularity is once again on the increase.

There was never any real technique talked about until the 1970s and 1980s when people like Alan Langton, Geoff Cookson and Chris Browne started talking about 'snapper snooping' or groundbaiting. These are very different techniques, but practiced in a similar depth.

Snapper love to feed in the shallows, but will also rest up away from predators and can be found in gutters, lying between boulders, or even just in kelpy bowls lying motionless or slowly circling in reasonably shallow water. This is when spearfishermen stand their best chance of catching one. Of course there have been many large snapper taken in deep water as well, but in the shallows the diver can approach slowly along the bottom as they have more bottom time, and can back off unseen and have another go should the first attempt not work.

The northeastern half of the North Island has the best snapper numbers and the best coastline to hunt them. The west coast is not as good because of heavy swells, and few of the islands that are necessary to provide the shallow, sheltered water snapper like to rest in.

'Snapper snooping' involves finding a sheltered, quiet piece of coastline with plenty of big boulders and deep gutters with lots of kelp and plenty of ledges to peer over. Divers are dropped off by a boat which anchors well away: it is also important that the boat doesn't travel too close to the shore, as this will scare off the fish.

The spearfisher should always swim with the sun behind them so as to blend into the surroundings as much as possible, as it is hard for a snapper looking into the sun to distinguish between a diver and one of the many stingrays found in these areas. The trick is to work along the coast no deeper than 10–12 m at a steady pace and check out every likely-looking spot.

Sometimes it is hard to stop and employ the ground-baiting technique for snapper, since divers feel tempted to swim on along a large piece of coast, drawn by the thrill of what may be over the next rock or around the next bend. But sometimes, owing to the disturbance caused by other divers, boats, or lack of good coastline, it is necessary to groundbait to get a snapper.

The diver needs to look for a point or rock that has a deep drop-off to open water with a little current flowing past, and a good approach that a feeding fish can't observe. Once a diver has been in a bay, every snapper in that area knows they are there, whereas a rock or point generally has access to deeper water where unsuspecting snapper will be holding or moving through.

Current flow helps distribute berley scent. A drop-off should not be much more than 2.5 m so a spearfisher can shoot at a fish without showing themselves. They can get away with sliding their gun over an edge towards a feeding snapper, as long as it is done quickly but without sudden movements. Sometimes, snapper will swim towards the

speargun. If the spear's folding barb is hanging down it can reflect the light and look like a baitfish in trouble. Some divers actually give their gun a little wiggle in this situation to excite the fish.

Kina (sea urchins) are usually preferred for berley, as there is never a shortage of them and they are part of a snapper's staple diet. If a snapper is seen in the open without being alerted, but cannot be approached without being spooked, the diver will pick up a couple of kina, crack them and, staying in cover, throw them so the snapper just sees these tasty morsels falling.

An alternative is to gather a stockpile of kina in one spot, crush them with a rock, then leave the area for about ten minutes, allowing the fish to gather around the bait. It doesn't pay to hurry, as big snapper will sometimes circle an area before going in to feed, and see the diver first. There is no hope of getting these ones, but persistence may yet pay off as generally others will turn up.

Once the snapper start feeding it pays to leave them to it for a little bit as they are generally on edge for a little while. The spearfisher has to let them feed and settle in before taking a shot, and be wary of the little ones that spot the diver and bolt, warning all the others. If this happens, the trick is to back off and leave the bait alone for a few minutes before returning.

If a fish is killed instantly when speared it may be possible to drag it away without disturbing the others. Then it may be possible to shoot several more before they become wary and will not feed off that bait again.

The best time to shoot a fish on a bait is as it comes in to feed, as this is when it is focused on its target and not the area around it. The hardest time is as they are leaving a bait; they seem to be on high alert and are easily spooked at this stage. It is only worth working a bait for half an hour before moving on and trying another spot.

Snapper are a very rewarding fish to hunt, and if a spearfisher gets a couple of nice fish for a whole day's effort, they can feel they have really achieved something: outsmarting one of the smartest and best-eating fish in New Zealand.

An iconic fish like the snapper has developed its own suite of art. This painting by Georgie MacBrayne (www.fish-paintings.co.nz) is titled Snapper and mussels.

JULIE-ANNE CUNNINGHAM

PORT WAIKATO TO CAPE EGMONT

09

Like the northwest coast described in Chapter 6, snapper habitat from south of the Manukau to Cape Egmont has limited road access and receives a lot of heavy weather, especially in the colder months, which restricts access. There are long stretches of black ironsand beaches that are not easy to get to, and much of the region is sparsely settled. The main access points are Port Waikato, Raglan, Aotea and Kawhia, then a series of small tidal river mouths such as Marokopa, Awakino, Mokau, Tongaporutu, Urenui and Waitara, before the major port of New Plymouth. However, these natural defences have not protected the region from the excesses of commercial fishing.

A HISTORY OF OVER-EXPLOITATION

The snapper fishery zone extending the length of the North Island's west coast, known as SNA8, shares the usual long, sad history of over-exploitation. As far back as 1953, trawlers, many from Onehunga, were taking over a thousand tonnes of snapper a year from the zone. This amount was sustainable, and kept around that level by the smaller size of the boats in those days, the few ports and unreliable weather. However, the Japanese started fishing this area in the 1950s, and although they were longlining all around the North Island, the sparse population on the west coast meant there was less observation of their activities by the NZ public. This meant the Japanese were more frequently able to infringe the closed fishing zones inside which snapper were most abundant.

It will probably never be known how many snapper Japan took from NZ waters before they finally withdrew in the early 1970s, but in 1964 alone 25 longliners made 48 voyages to NZ, catching 5800 tonnes of snapper. This doesn't count the trawler catches, and obviously not all of this was from SNA8.

Even after the Japanese left there was little respite. By 1973 the SNA8 domestic catch had crept up over 3000 tonnes, and by 1976 reached a reported peak of 5326 tonnes (though the catch was considerably under-reported, so the actual figure would have been significantly higher). The fishery could not sustain this pressure and started a long slide down to only 893 tonnes in 1986, the year it was brought into the Quota Management System (QMS).

Even the QMS wasn't the answer. The Total Allowable Commercial Catch (TACC) in SNA8 was originally set at 1330 tonnes to allow the stock to rebuild, but was progressively raised to 1594 tonnes by 1989 as a result of extra quota issued by the Quota Appeals Authority (QAA). It was finally set at 1500 tonnes for the 1992–3 season, but between then and the 2004–5 season the deemed value scheme (see Chapter 4) added a further 1234 tonnes. Small wonder that this fishery has been so slow to rebuild, and especially since the area lacks the ideal spawning conditions of the east coast.

In 2005 a new stock assessment estimated the snapper population to still be only 50% of the biomass that would provide Maximum Sustainable Yield (MSY), and only 8–12% of virgin stocks. The Ministry of Fisheries proposed a range of catch reductions to help replenish stocks. The commercial industry rejected any cuts to its catch, and

OPPOSITE , CLOCKWISE FROM TOP LEFT: A surfcaster tries his luck in the sheltered Raglan Harbour, where plentiful shellfish are an attraction for fish and human alike.
SAM MOSSMAN

Modern Kiwi kite-fishing pioneer and recreational fishing advocate Paul Barnes demonstrates his methods on a blustery West Coast day.
COURTESY PAUL BARNES

For boaties, the West Coast has few chinks in its armour. The Raglan bar is one, but to be approached with considerable caution and with the benefit of local advice.
SAM MOSSMAN

BOTTOM LEFT: Paul Barnes with a pair of kite-caught west coast snapper.
COURTESY PAUL BARNES

The concept of a propeller-driven, torpedo-like device has been around for a while, as shown by these undated but evidently old photos sent to *NZ Fishing News* magazine by Auckland Kon-Tiki fisher Neville King in 1998. The heavy steel submarine with snorkel and exhaust was powered by a small single-cylinder petrol engine driving a 'puller' propeller. It was reputed to be one of three constructed in the Wellington Ministry of Works workshop.

recreational fishers supported the greatest of the proposed cuts to commercial quota, but argued against reductions to recreational limits, saying that it was commercial overfishing that had caused the damage.

In the end, the Minister of Fisheries decided that all users should share the pain of rebuilding the stock. The total catch was reduced by 275 tonnes, and recreational limits were cut from 15 to 10 snapper per person per day in the northern region. Importantly, the deemed value of snapper was also increased to discourage 'accidental' commercial catches.

FISHING THE WEST COAST

Despite the problems in SNA8, recreational anglers fishing this coast can still have some exceptional experiences. Both private and charter boats from Raglan, Aotea and Kawhia get some fabulous snapper action off this section of coast in the spring months – if and when the weather lets them cross the notorious harbour bars. Likewise, owners of small boats can launch (with caution) from river mouths such as the Awakino, Mokau, Tongaporutu, Urenui and Waitara. Generally, boating off the west coast is much more demanding than on the east coast with its many sheltered ramps, harbours and bays.

Port Taranaki at New Plymouth is the best boating access to the southern part of the region, with a sheltered ramp in the harbour and an active boating and fishing club. Further south is the Cape Egmont Club's facility.

BY HOOK OR BY CROOK

It would be strange if the famous 'Kiwi ingenuity' was not applied to catching snapper. Over the years, Kiwi fishermen have developed and refined many ways to run longlines out from the shore.

Shutterboards (also called 'Galloping Gerties' or 'bottom creepers') are a system that uses wave-power to get a line offshore. A weighted bottom board is fitted with a hinged flap or shutter which folds flat under the pressure of incoming waves, but lifts and presents resistance to the backwash, which drives it out to sea. However, it is slow and heavy to retrieve.

GO FLY A KITE

The use of kites in fishing is recorded by early Pacific Island cultures. American charter skippers used kites to catch shy bluefin tuna from boats in the waters around Catalina Island as early as 1905. The British media reported successful kite fishing by a shore-based angler in 1901. Kites have become a popular way of running longlines from the shore in recent years, and this is not the first kite-fishing boom NZ has experienced.

Ray Doogue, in his 1967 book *Hook, Line and Sinker*, describes successful kite fishing from the shore in 1955. He notes that he was far from alone in this venture, and a press report of the time mentions aircraft pilots complaining about the dangers of fishermen's kites on the Kapiti coast in the 1950s. In a later book *Saltwater Fisherman* (1976) Doogue describes in detail the kite-fishing technique of a Hamilton-based angler named Osborne, a technique which was the forerunner of the successful techniques used today.

Paul Barnes, of Paul's Fishing Kites, is the doyen of modern kite fishing, and today kite-fishers are common, especially on the west coast of the North Island. Barnes, an ex-commercial fisherman, developed his first kite-powered longline in 1987, having left the industry and sold his boat.

He remembered the frustrations of fishing Kon-Tikis and 'windbags', and after reading in a library book that winds were more constant and steady higher above ground, decided that towing a longline with a kite might be the answer. Using his commercial longlining experience, he built a kite rig that was so successful on its debut at Muriwai Beach (catching over 80 gurnard in a day) that a flood of requests for similar devices set him on the path to a successful business.

After a couple of years, losing gear to sharks and snags on the west coast prompted him to develop the sacrificial dropper-rig, allowing the kite rig to be retrieved with only the loss of terminal tackle. This rig also doubled the number of fishable days by negating the effects of rips and surf. Further development of kite-fishing technology has produced scaled-down systems that can be used to tow out a smaller rig from a rod and reel.

'We can tack out kites at 60° to the beach, and with two kites can fish successfully in conditions ranging from four to 60 knots of wind. The whole philosophy behind kite fishing is to balance the technique with the available wind,' says Barnes.

Having developed a wide range of kite-fishing accessories, Barnes now exports his fishing kites around the world. Research and development continues: he is testing helium-assisted kites he calls 'Kytoons' and trolling kites for game fishing. His kites have even been used for sea-bed profiling in Antarctica. He has also been a tireless worker as project leader for option4, the recreational fishing rights group.

During the spring months, some very large snapper approach the shore in parts of the country. One such region is Mokau, between Kawhia and New Plymouth. This region is legendary among kite fishers, but because of snags and current it is not suitable for fishing with electric Kon-Tikis. The area even has a professional kite-fishing guide, Geoff Preston. Geoff has caught many snapper in the area weighing double-figures, and his biggest (caught in October 2006) was a magnificent 15.7 kg (34.7 lb) fish.

THE KON-TIKI

The sailing raft is another venerable shore-based longlining technique, and is universally known as the Kon-Tiki, after the famous balsa-wood raft Thor Heyerdahl sailed from Peru to Polynesia in 1947. Kon-Tikis were popularised by Gainor W. Jackson in 1962 in

BOTTOM: The Kon-Tiki was popularised by Gainor W. Jackson in 1962 in *The New Zealand Beach and Boating Book*. It featured an ingenious Heath Robinson-style timing device to control the distance it travelled out to sea. A boiled sweet immersed in the water took about 20 minutes to dissolve, causing the string that held up the sail to come loose. As the sail fell, the string also slipped through a sinker and released it to slide back down the line, anchoring the raft and pulling the baited line to the bottom where the fish were.

BELOW: This electric Kon-Tiki, built by Neville King, is of the familiar raft style, but powered by a 12-volt sealed lead/acid battery which ran a small electric motor turning a propeller on a stainless steel shaft. Wheels under the pontoons aid beach handling. *Grant Dixon, NZ Fishing News*

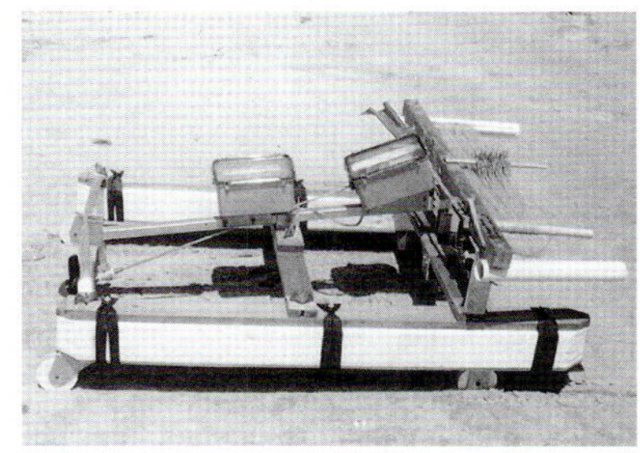

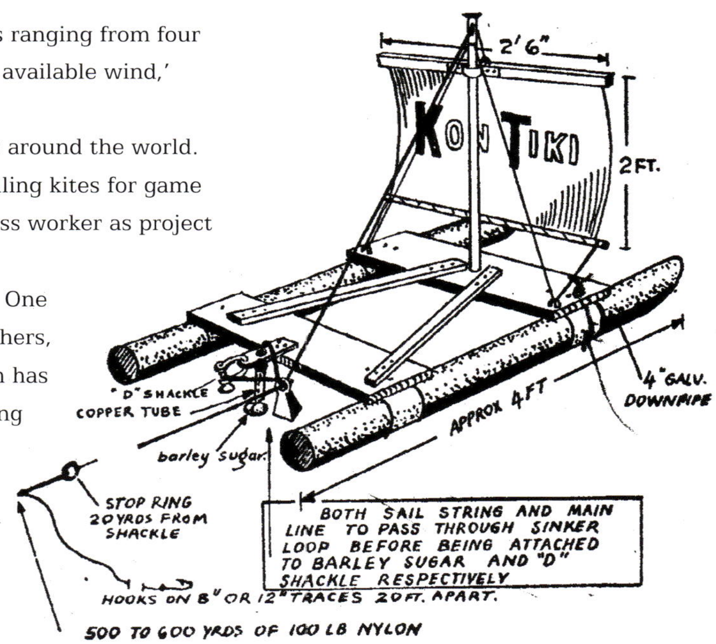

BELOW: Justin Fox (left) and Geoff Preston with 10 kg and 13.6 kg snapper caught from the black sands of Mokau, a renowned big-snapper area for west-coast kite fishing. *COURTESY OF GEOFF PRESTON*

BOTTOM: Ray Hughes with his 'twin-hulled' electric Kon-Tiki, built in 1998. *GRANT DIXON*

The New Zealand Beach and Boating Book. This design featured a simple timing device to control the distance it travelled out to sea – the epitome of Kiwi ingenuity. A Heath Robinson-style timing device used a barley sugar or similar boiled lolly immersed in the water. This took about 20 minutes to dissolve, dropping the sail and releasing a sinker to slide back down the line, pulling the baited line to the bottom where the fish were and anchoring the raft.

Ted Bradstock built one in 1963 and used it for some years to catch snapper and other fish from the beach in front of his bach at Whangapoua, on the Coromandel Peninsula:

'I soldered up the pontoons from scrap lengths of downpiping, the wooden frame was packing-case oregon and the sail was an old pillowcase. The reel was made from scrap wood, number-eight wire and a bit of old pipe. The only parts I had to buy were the line, clips, swivels and hooks.

'Whenever an offshore wind sprang up we'd bait up 20 hooks and get it out smartly. If the wind was strong the Kon-Tiki took rather less than 20 minutes to go out so you'd suck the barley sugar a bit first to make it smaller.

'We got some pretty big snapper in those early days, including a 14-pounder, and sometimes found broken traces which had obviously been broken by bigger fish.

'For a time everyone seemed to have a Kon-Tiki, but the catch rate plummeted as the commercial fishery grew. Often at night you could hear trawlers chugging along close in to shore and they soon cleaned out the snapper.'

GOING ELECTRIC

However, the Kon-Tiki raft (or its simple cousin the airbag) has limitations. Use can be restricted by wind direction and strength, or surf which can pick up the raft and dump the whole thing back at your feet.

Just as engines took the place of sail in commercial shipping, so it was with Kon-Tikis, as inventive anglers worked out ways to break their dependence on wind direction and surf conditions. Initially, a series of one-off devices were built by amateurs. Back in the late 1980s an Auckland fisherman and model aircraft enthusiast called Peter Johnson built a twin-pontoon device from spare model aircraft parts, a 16-cc two-stroke weed-eater engine and a few other bits and pieces. His Kon-Tiki was remote-controlled and had a servo device to drop the sinker in response to a radio signal.

Around the same time, Bill Homes of Papamoa adapted his traditional Kon-Tiki with a Briggs and Stratton motor mounted on its side and a car-radiator fan to make kind of 'air boat'. Bill used the traditional barley-sugar method for a tackle release and regulated the run-out time by the amount of petrol he put in the tank.

These were many backyard devices used to set longlines from the shore. They worked, were good fun, and to a certain extent beat the dependence on wind. However, they were not immune to being overturned by waves or soaked by water. Designs took a new tack – the electric Kon-Tiki.

The concept of a propeller-driven, torpedo-like device has been around for a while, but advances in materials,

THE SNAPPER KING

The limited boat access has made shore fishing very popular here. Many talented surfcasters have come from this area, such as Leon Jury. This softly-spoken, unassuming surfcaster hails from New Plymouth. Wherever anglers gather he is recognised as the best, having won more national open surfcasting titles than any other angler. He has beaten the 20-pound benchmark on 78 occasions in New Plymouth waters alone.

Leon's first snapper came from the New Plymouth wharves where he fished with his dad as a boy. He and another lad absconded from a school trip to the nearby port. After a short time, his friend left to sneak back to the school party. Shortly after, Leon hooked his first-ever snapper on a light herring line, and after a long struggle brought the 6 kg fish to the surface. For a full hour Leon had the fish lolling on the top and was trying to find a way to land it. The light line eventually broke, leaving Leon with a moral victory and just enough time to rejoin his class mates, unmissed.

Master snapper catcher Leon Jury holds a 24 pound snapper which has the body of a 30 pounder. Big snapper like this one have been a constant surfcasting catch for Leon. He caught this fish on the day he saw the biggest snapper of his life, which he says 'made this fish look like a baby'. *Jim O'Brien*

A profound influence on Leon's surfcasting life was Bob Thompson, a huge and likeable man whom Leon rates as the best surfcaster he ever saw. Bob was also a New Zealand dry-land casting champion, and asked Leon if he would like to join the New Plymouth Surfcasting Club. This he did in 1958 and Bob, along with another well-known Taranaki fisherman, Noel Baty, became his firm friends. Bob and Leon fished everywhere on the huge Taranaki coastline, discovering new spots weekly and finding the best way to fish them.

After fishing widely Leon still believes the Taranaki province offers NZ's best surfcasting, and he has yet to fish it all. Regardless of the bait used, Leon is adamant that the most important part to snapper fishing is to present the bait in the right place and at the right time.

Leon's first 20-pounder came from the 'Kumara Patch' at the mouth of the Stoney River while fishing with his friend Bob, who also caught a 20-pounder there. This was the weight after the fish had been gutted, as was the custom of their club at the time.

Ask about the best day's fishing Leon ever had and his voice adopts a rather melancholy note as he tells of a day he and Bob fished the Tapawai Reserve. Access was by walking over paddocks and descending the cliffs via a thick rope. The fishing was unbelievable, and among their catch were four snapper over 20 pounds. Sadly, Bob died the next day, and Leon still wonders if the arduous climb back carrying their gear and fish was a factor. There would be few anglers though that would have had a last day's fishing as good.

Leon is still striving for his first 30-pounder, but his best effort of 29.75 pounds still isn't too bad. The biggest snapper he ever saw was during the national surfcasting champs near Te Kaha in the Bay of Plenty. Because of a faulty freezer Leon had to unload his large stash of berley into a very fishy-looking rock gutter. He managed a big but skinny snapper of 24 pounds. With the fishing now quiet Leon made his last cast, then began flicking little bits of residual berley into the water. As if from nowhere, the biggest snapper Leon had ever seen began chomping into the scraps, its massive back broaching out of the water.

Shaking with excitement Leon retrieved his gear, cut the sinker off and lobbed his bait within a metre of the fish. The snapper circled the bait a couple of times ... and then disappeared into the open sea.

technology and manufacturing processes allowed the evolution of electric Kon-Tikis with semi-submersible, fully-sealed designs featuring internal batteries powering drive units originally made for electric outboards. These were a great success and although expensive compared with kites, a commercial demand was established, especially from retired folk or those who could not afford a boat.

Opotiki-based Chris Southern came up with the Robot Fishing Mate in 1993; Ray Hughes from Tuakau released his Electric Kontiki in 1998, and there was Hamilton-based Colin Gatland's 2002 Powertiki, the Kentiki, and the Seahorse, to name some of the products marketed commercially.

Although longlining does not give the thrill of feeling a bite, followed by hooking and fighting a fish on a rod, there is no denying that there is still a delicious anticipation in pulling a longline in – like unwrapping a fishy Christmas present.

THE 'BIG BANG' THEORY

Like a small boy's fascination with fireworks, there were big boys who considered the best and quickest way of getting a longline out to sea was by the use of ordnance. A lot of this experimentation was a bit clandestine: there was some doubt as to the attitude of 'the authorities' to the use of artillery on the beach.

It is rumoured that some early devices involved the use of a galvanised water pipe and reloading powder, but there were problems with rapid acceleration ripping off the baits or breaking the line. More recently came the development of the 'spud cannon' – a PVC pipe device powered by ether-based spray-can propellant and set off with a barbecue igniter. It could shoot a projectile (half a potato) towing the fishing tackle out about 500 metres.

TOP: While kites go over the surf, electric Kon-Tikis go through it, regardless of wind direction. Mike Finnigan launches his Seahorse through the west coast breakers. *SAM MOSSMAN*

ABOVE: Arthur Mayall and Heather Baker with the results of a single set of their electric Kon-Tiki. *SAM MOSSMAN*

JULIE-ANNE CUNNINGHAM

THE EAST COAST AND HAWKE BAY

10

At the core of the region stretching from East Cape to the southern extremity of the Wairarapa coast, lies Hawke Bay, bounded by Mahia Peninsula and Portland Island in the north, and the long grey/white finger of Cape Kidnappers in the south.

This has always been a prime area for snapper – a huge basin with a rich, mostly sedimentary bottom, supporting large amounts of snapper forage. The massive stable water mass of the Bay provides ideal spawning grounds, 30–50 m deep, and while its more southerly position means it does not maintain the preferred spawning temperatures of around 18°C or more for the length of time that the more northerly snapper Valhallas do, January and February used to see large spawning congregations of these fish in its mid-depths. Fishermen now in their seventies and eighties recall seeing surface snapper schools giving a red/brown tint to the sea off the beach in the East Clive area.

Hawke Bay used to have a massive snapper nursery and rich inshore feeding ground, in what was subsequently called the Ahuriri Lagoon. Visitors driving in from the north into modern-day Napier, or travelling in from the airport, should not be taken in by the small area of wetland and estuarine area they now cross. When Cook first sighted the region in 1769 he described it as 'a bluff head … on each side a narrow stone or sand beach … between these beaches and the mainland is a pretty large lake … of saltwater as I suppose.'

The 'bluff head' became known as Scinde Island and was the home of early European settlement in the region. The massive Ahuriri Lagoon and all of the flat land for a considerable distance to the north and south of it were then either under water or at least tidal marshlands. This rich ecosystem was nibbled away by reclamations over the early years of settlement, but the major change came with the Napier earthquake of 1931, which not only levelled much of the town, but raised the land by 2.5 m, draining much of the estuary.

Where the lagoon once was now stands Napier Airport and much of the Napier urban area. The loss of this huge nursery and feeding area and the nutrients it had fed into the Bay must have adversely affected the fish life in the area.

Hawke Bay is still prime snapper habitat, with nutrient input of from five major rivers (the Wairoa, Mohaka, Tutaekuri, Ngaruroro and Tukituki) plus many lesser waterways. These rivers bring with them the worn river-stones which can be seen on many of the beaches in the central and southern regions of the Bay, and form a number of boulder-bank reefs; but generally there is a paucity of reef on the seafloor of Hawke Bay.

COMMERCIAL EXCESS

Napier was one of the early centres for the development of commercial fishing, particularly trawling. Hawke Bay was ideal for this fishing method, with few natural obstructions for nets to snag on, and large areas of moderate 20–50 m depths which were easy to work. Fishing out of Napier quickly recovered after the 1931 earthquake, and between

The long grey/white finger of Cape Kidnappers marks the southern end of Hawke Bay and more or less the southern extent of reasonable stocks of spawning snapper on the east coast of the North Island. *SAM MOSSMAN*

OPPOSITE TOP: Prior to the Napier earthquake of 1931 the massive Ahuriri estuary was a pathway for nutrients into Hawke Bay waters, and provided an extensive nursery area for fish, including snapper.
COLLECTION OF HAWKE'S BAY CULTURAL TRUST – HAWKE'S BAY MUSEUM, NAPIER, NZ 8004

BOTTOM RIGHT: Fishermen mend and dry their nets on the beach in the 1920s. The young boy with them is Peter Tait, who became an MP and Mayor of Napier from 1965–1972.
LESLIE HINGE COLLECTION, ALEXANDER TURNBULL LIBRARY, WELLINGTON, NZ. 1/2-034777-G

BOTTOM LEFT: Hawke Bay has a long history of commercial fishing. This early photo shows the trawler fleet tied up in Port Ahuriri.
COLLECTION OF HAWKE'S BAY CULTURAL TRUST – HAWKE'S BAY MUSEUM, NAPIER, NZ 4663

1934 and the beginning of the World War II snapper landings in the region were seldom under 100 tonnes.

After a break from serious commercial fishing, during and following the war (bottoming out at 53 tonnes in 1948), commercial catches increased through the 1960s, peaking at 856 tonnes in 1967. For the next decade commercial landings in this snapper fishing area (known today as SNA2, extending from East Cape to Kaikoura) remained very high – over 700 tonnes – and peaked at 878 tonnes in 1972. These unsustainable levels were maintained by the introduction of pair-trawling in the 1970s: two boats towing a single outsized trawl net between them.

Pair-trawlers could scoop up schools of snapper as they came together for spawning in late spring and summer. However, wastage was rife: in 1978 one operation took a massive catch over the Christmas break, only to find the processing plants were closed for the holidays. Seven truckloads of prime breeding snapper had to be buried at the local tip. After such excesses it was little wonder that commercial landings crashed to 374 tonnes in 1977, and by 1982 were down to 160 tonnes.

Snapper populations were not helped by the influx of Japanese commercial fishing boats that first came into NZ waters in the late 1950s. Hawke Bay anglers often found their longlines in the late 1960s and early 1970s. One irate boatie passed along several miles of line, smashing its glass floats. Others souvenired sections of the lines, encountering the first recurved hooks they had seen. They were not slow to copy this technology, and were using recurved hooks to great effect by the early 1970s.

Oil exploration in the 1960s and 1970s in the East Coast Basin also had a detrimental effect on the snapper population. Grid surveys involving the regular detonation of seismic charges were carried out. Boats in the areas being surveyed reported big fish kills from the explosions, and people walking the beaches often saw dead fish and the remains of the orange plastic-bag floats with cables attached.

DARK DAYS FOR SNAPPER

The period from the late 1970s until the turn of the new millennium was a dark time for recreational fishermen in Hawke Bay. The once-plentiful snapper were virtually gone, and an unbalanced ecosystem see-sawed with population explosions of undesirable species (for anglers), such as spiny dogfish and red cod which temporarily expanded into the vacant feeding niche.

Something had to be done, and finally in 1986 the Quota Management System (QMS) was introduced. A closed fishing zone called the 'Wairoa Hard' in the north of Hawke Bay gave snapper populations an area from which to slowly replenish. It was a slow business, not helped by the 'deemed value' system, which allows commercial fishermen who accidentally catch a species for which they do not have quota, to keep and sell the fish after paying a penalty – the 'deemed value' of the fish. Deemed value is intended as a management tool to avoid wastage, but creates problems of its own because sometime the fish are worth more than the deemed value the fisherman pays for

them, and it becomes tempting to catch them 'accidentally-on-purpose'. This merely trades one problem – wastage – for an even bigger problem: overfishing.

In every season but one between 1987 and 2005, the quota for SNA2 has been exceeded, often by a large amount: in 1990–91 the permitted catch was 157 tonnes, but the actual catch was 427 tonnes, mostly reported as a bycatch from the tarakihi and gurnard fishery. Although the commercial quota for snapper has subsequently been increased stepwise to 315 tonnes since 2002–3, the commercial bycatch has nearly always remained well above the quota set to allow the fishery to recover.

Readers may draw their own conclusions, but over the 18-year period, more than 1,500 tonnes of snapper have been taken in excess of the quota. Assuming an average weight of 2 kg per fish, that is an excess of 750,000 snapper caught commercially above the quota. No wonder it has taken nearly 25 years for snapper to make some sort of comeback.

Northern Hawke Bay was the first area to experience reasonable snapper fishing again, followed by the central part, and since about 2002 fishermen in the south have started to see these fish regularly again, although the numbers are still low. Schools of large snapper used to be encountered in high summer on the spot known as 'Razorback Line' off Cape Kidnappers, and even to the south of this, off Whitecliffs and towards Waimarama. This seems to be the edge of the range for Hawke Bay snapper, and populations have not yet built up large numbers in the southern zone.

NORTH AND SOUTH

From Hawke Bay (the core zone of the east coast snapper) the coastline runs north to the natural barrier of East Cape. There is a wide variety of foul ground along the coast and offshore that provides good habitat and feeding grounds for snapper. However, these populations are constrained by the more limited spawning opportunities, and anglers in this region are just as likely to pursue tarakihi or groper (hapuku) as snapper.

Commercial fishermen claim that snapper numbers are slowly increasing on this section of the coast, and this may be because of fish that have come round from the Bay of Plenty and north from Hawke Bay. At the very top of this beat East Island with its inshore reefs, and the offshore Ranfurly Banks, can provide some good snapper fishing.

South of Hawke Bay is the wild and relatively inaccessible Wairarapa coast. Spawning opportunities are even more limited here, but long-time anglers have reported the occasional large 'moocher' in the past. Possibly these large fish are loners that have strayed south for a more solitary lifestyle along the rocky, food-rich coastline. John Pothan of Porangahau reports:

'I can recall, at low tide, seeing holes left behind where the snapper had been in feeding on pipis. Sometimes you could see them on their sides, feeding in the shallows. Before 1965 we had a good supply of large snapper – on one

DO HOOKED SNAPPER SURVIVE RELEASE?

In early 2008, the minimum legal size of recreationally-caught snapper was 27 cm, a size calculated to allow them to breed at least once before they could be kept when caught. In practice, many anglers consider this to be too small and use 30 cm as their personal minimum. Anglers are also often encouraged by the fishing media to release the 'big breeders'.

Consider: a 27 cm fish is just coming up to its second spawning and could release more than 100,000 eggs over the spawning season. In comparison, big fish each produce millions of eggs, and they have also proved their fitness and will pass their proven genetically-based growth and survival traits on to their offspring. By killing the biggest fish and releasing small ones, humans may well be selectively breeding for smaller or inferior snapper.

Research shows that 96 percent of mouth-hooked snapper will survive being caught and put back. *CRAIG WORTHINGTON*

So, do snapper survive being released, and what factors affect the survival rate? In 1995 Jeremy McKenzie and John Holdsworth carried out a field investigation into the mortality rate for released line-caught snapper in the inner Hauraki Gulf.

They got one group of recreational anglers to catch and handle snapper in ways that may be expected in a typical fishing trip, and a second group to handle their fish with the utmost care. The latter group were instructed to lift the fish carefully from the water and place them on a wet foam pad, wet their hands before touching the fish, remove the hook carefully, and not to touch the gills or eyes, squeeze the gut or gill plate or lift fish by their tails. If the hook was set any deeper than the jaw, they were to cut the trace and leave the hook rather than risk causing more damage by pulling on it.

These fish were released into a large sea cage for 15 days, and any dead fish removed twice daily by divers. The survivors were then removed and examined. Fish that showed signs of two or more types of damage were counted as mortalities.

Only 4% of mouth-hooked snapper died after release, while the rate was about 75% in fish hooked in the throat, gills or gut. The clear conclusion was that the key factor in fish mortality after release was where and how they were hooked.

Another issue was how handling affected survival. The study could find no differences in survival rate between 'normally' handled fish andd those which were carefully handled. It must be remembered, however, that these fish were all small (under 35 cm), corresponding to the small fish that are usually released by anglers. Small fish are light, so there is less strain on their bodies when out of the water, and they are easier to handle without suffering adverse effects.

Larger, heavier specimens need some body support if handled when out of the water. It is probably best to avoid putting them on dry surfaces: using a wet rag or old towel is a useful way of holding a fish for unhooking, as it will give you a good grip, avoid injury from spines, scales and gill plates, and also avoid damaging the fish. Efficient handling gets the fish back in the water quickly – an important part of reducing the mortality rate. If a fish is deeply hooked, it is best to just cut the line as short as possible and leave it – but remember: deep-hooked fish are much less likely to survive, so if they are of legal size, they are best kept for the table.

To avoid deep-hooking snapper don't use small hooks: 4/0 should be a minimum, with 6/0 or larger preferred. Circle (recurve) hooks increase the chance of jaw hook-ups.

trip out to the beach with a friend in November 1965, I caught five snapper all over 25 lb, with the last fish breaking my cane rod.Today, sadly, we do not see numbers of snapper south of Hawke Bay anymore. They are very slowly coming back but trawl activity keeps them down. There is the odd good one though. I run a competition each year for the largest snapper to be weighted in at Porangahau; in 2005 the winner weighted 3.68 kg; in 2006 it was 8.36 kg, and so far in 2007 the leading fish is 10.42 kg.'

TOP: In spring large snapper move in along Hawke Bay beaches to feed on paddle crabs, an event eagerly awaited by shore-based anglers. John Fulton caught this 16 kg beauty on his electric 'Kentiki' at Gills Road, just north of Napier Bluff.
BRUCE BASHER

ABOVE: Snapper have made a modest comeback in Hawke Bay, 30 years after heavy commercial pressure from trawlers targeting spawning aggregations nearly wiped them out. Rex Mossman with a fine fish taken off Napier in January 2007.
SAM MOSSMAN

SNAPPER IN THE BAY

Keen surf fishermen look forward to a spring run of pre-spawning giant snapper that move into the surf zone, starting in August and September. These big fish, sometimes exceeding 14 kg, are not exactly thick on the ground, but each year surfcasters, inshore boaties and Kon-Tiki fishermen pull in some huge specimens from the beaches close to Napier. It seems that these XOS fish move in from the deep water, where they spend the winter, to feed on the paddle crabs common along the beaches, as these are the usual stomach contents of the fish caught. For surfcasters these monster snapper are the fish of dreams, and the dedicated will sit for many hours even in the rain, cold and dark, on the offchance of catching one.

Later in the spring, school snapper put in an appearance, often moving from the northern zone into the central and coastal area. Smaller fish can be found in summer around underwater structures such as the inshore reefs and boulder banks around the coast, a few wrecks, and the oil-exploration borehole Hawke Bay-1 and its associated anchor blocks. Fishing shallow inshore grounds will often produce good numbers of school snapper.

As spawning approaches, the schools of mixed-sized fish congregate at a depth of about 35 m in the area between Napier and the freshwater springs offshore about 35 m deep. This is mostly an area of open sedimentary bottom, and successful fishing is a matter of anticipating the movement of the schools, then anchoring and fishing the bottom. Ledger and flasher rigs are commonly used, but large straylined baits cast well out and allowed to drift towards the bottom sometimes yield very large snapper. During the height of this activity (typically from Christmas to February) the best areas can be identified by the clusters of boats anchored there, and of course the commercial boats, which are out for their share too.

TAKING TO THE WATER

Much of the boat-fishing effort is based out of Port Ahuriri, the base for the Hawke's Bay Sportfishing Club and the Napier Sailing Club, as well as the location of a marina and other facilities.

To the south, at Clifton Beach is another popular fishing base with launching facilities run by the Clifton Marine Club. Here, launching and retrieving is over a beach with changeable conditions, and the locals have developed a unique system. Boat and trailer (with floatation built in) are launched into the surf and driven off the shore with the outboard. The trailer is then anchored, the boat detached and driven off for the fishing trip. On its return, the boat is driven back on the trailer, cinched down, and driven back up the beach with the outboard, then pulled clear of the water with one of the club's winches. A great example of Kiwi ingenuity solving a launching problem.

The Gisborne-Tatapuri Sportfishing Club has one ramp in Gisborne central and a second to the north at Tatapouri Beach. And everywhere from East Cape to Cook Strait, are spots where people launch off the rocks, beaches or out of river-mouths, often with the aid of quad bikes, 4×4s or the traditional beach tractor.

THE PRESENT

It is fair to say that snapper in Hawke Bay have made a comeback since their near-extermination in the early 1980s. Indeed, modern anglers with no knowledge of how this fishery used to be might think that the fishing today is fairly good. This is an effect called the 'sliding baseline' – each generation can only make comparisons to their own experiences. Hawke Bay snapper still have a fair way to go, especially in re-establishing their southern populations. This is not helped by commercial boats overfishing the quota, and fisheries ministers who increase the quota at the first sign the stock is making some progress in rebuilding.

BOTTOM LEFT: Only trawlers up to 13 m long are allowed to work the area off Napier that holds spawning snapper during the summer months. The *Thelma G* works her nets among the recreational fleet off Napier in January 2008. SAM MOSSMAN

BOTTOM CENTRE: An answer to the problem of launching a boat on an exposed coast. The boaties at Clifton Beach add flotation to their trailers, strap the boat down and drive the whole assembly out to sea, anchoring the trailer and driving the boat off it to go away fishing, and reversing the procedure on their return. A winch is used to pull the rig up the beach. SAM MOSSMAN

BOTTOM RIGHT: At Waimarama Beach, south of Cape Kidnappers, the faithful old beach tractor and partial shelter of a small reef allow boaties to beach-launch their (mostly) pontoon boats. SAM MOSSMAN

JULIE-ANNE CUNNINGHAM

WANGANUI AND WELLINGTON

11

The stretch of coastline south from Patea to Kapiti Island consists of over 150 km of long sandy beaches interspersed with river mouths. Some major rivers enter the sea along this coast, including the Wanganui, Rangitikei and Manawatu, and their often unpredictable bars restrict recreational boating and fishing to periods when the prevailing westerlies relent.

During the colder months, less-sought-after fish species like red cod and spiny dogfish predominate, but from October to January snapper are more common. Snapper numbers then decline, with only the odd one still caught after February.

This section of coast is largely sedimentary, but there are deeper sections and areas of low, rolling foul. The main one is a long stretch of patchy reef 8–12 km offshore running south from Waverley Beach to the end of Kapiti Island.

When conditions are right, small boats can launch from the beaches along the coast while the larger craft still use ramps. Wanganui and Tangimoana (the Rangitikei River mouth) are popular departure points, while the Manawatu at Foxton has a bar that can be dangerous. Further to the south, ramp facilities are available at Otaki, Paraparaumu and Waikanae. However, it is little wonder that with so many wide, open beaches, surfcasting and longlining with kites and Kon-Tikis (see Chapter 9) are popular ways to fish along this coast. The greatest concentration of anglers in this region is around Wellington.

A CAPITAL FISH

A national recreational fishing survey in 1986 reported 224,000 New Zealanders had caught a snapper, and a survey just one year later reported that anglers considered snapper to be their favourite species. For Wellington anglers this is certainly true. Since the early 1980s there has been a huge decline in the number of snapper caught from the shore, but boat-fishing for snapper is booming. This seems to suggest a decrease in the food supply, or degradation of the inshore habitat.

About the time successful surfcasting for snapper began its downward spiral in the Wellington area, boat-fishers realised that snapper were seasonally available and that Wellington Harbour was also a good snapper fishery. As techniques and timing improved, this wonderful fishery exploded and is going strong.

SNAPPER FROM THE SHORE

Early land-based snapper fishing was mostly in Wellington's bountiful harbour, which had good populations of fish. The strange thing is that whichever direction snapper enter the harbour from, it is very rare to catch them within 15 km outside of the entrance.

Records of snapper fishing pre-1950s are poor, but it appears that during this pre-rod-and-reel era, it was the dozen or so wharves that provided the opportunity for the handline experts to strut their stuff. Handline fishers would hold their lines or use a 'rattler': a tobacco or shoe-polish tin filled with small stones and attached to the line by a

TOP LEFT: An early snapper longline on a reel, probably in the 1950s. It was towed off Paekakariki Beach by the dinghy in the background (note spray dodger).
DIANNE HOOPER

ABOVE: Wellington's many reclamations have long served as fishing platforms. This shot taken about 1910 shows two men fishing with handlines off reclamations at Te Aro, near Cable Street. Either fishermen were better dressed in those days, or these men were on their lunch break. *THE PRESS (CHRISTCHURCH) COLLECTION, ALEXANDER TURNBULL LIBRARY, WELLINGTON, NZ. 1/1-017736-G*

LEFT: Early commercial fishing methods were labour intensive. Men of the Dellabarca family haul a beach seine at Lowry Bay around 1907.
ALEXANDER TURNBULL LIBRARY, WELLINGTON, NZ. PACOLL-7985-17

rubber band. If a fish ran out with the line it would noisily drag the tin and alert the angler. Back then, the weight of any snapper was measured in pounds and while most anglers didn't actually weigh their fish, there seemed to be few fish under 10 lb, and many of up to 20 lb. The bigger fish that were abundant in the earlier days seemed to be mostly caught from inner harbour wharves and in Evans Bay, where the waters were deeper.

One of the few written records from the early days of handline fishing for snapper comes from Geoff Clementson, who wrote about fishing from the 2 km wharf built during the Second World War between Eastbourne and Ward Island as harbour protection from submarines and torpedo boats.

'My father and I fished very successfully from this boom during the time of its existence. About three-quarters of the way out to Ward Island there was a large area of weed in which grew mussels, pipis and sea eggs, among other sea life. In this weed we caught plenty of snapper on cotton hand lines – no fancy rods and reels in those days. Early morning and evenings were the best times for fishing and we often went home with a sugar bag of fish.'

While young anglers had small baitcasting rods for catching kahawai in the early 1950s, it was the introduction of heavier nylon handlines that really opened up snapper fishing. A typical May school holiday trip for enthusiastic young anglers was to visit Burnham Wharf in Evans Bay. Using freshly-caught slimy mackerel for bait, they would throw the unweighted line out about 30 m, from where it would sink to the bottom close to the wharf's edge. Then the line was simply tied around the nearest bollard and checked every 10 minutes or so.

Returning to find the line stretched absolutely tight from the bollard was *Boys' Own* stuff, and often the fish was a huge snapper, made all the bigger by youthful excitement. Although fish were seldom weighed, the impression remains that many were around 20 lb. Surfcasting rods brought a different dimension to harbour snapper-fishing, enabling anglers to fish places where handlines were impractical.

NOSTALGIC TACKLE

The early surfcasting rods of the late 1940s were the classic split-cane rods with overhead reels, but it was during the early to mid-1950s that they really began to gain momentum. A 12-foot Rangoon cane rod with brass ferrules, fitted metal reel-seat, plastic foregrips and wire rod-guides cost the princely sum of ten shillings. The imported reels were relatively pricey, however. Overhead reels were popular, and the model of choice was the Penn Jigmaster, while the cheapest was the Penn Seaboy. Made of an early plastic and costing £3, these reels needed careful handling but cast very well. A 4 oz sinker could be regularly thrown over 130 yards using 19 lb line and no leader.

While those who had started fishing with the similar baitcasting reels chose to continue with overhead reels, newer anglers went for fixed-spool reels, usually Delfino and Mitchell. At the time, solid-glass surf rods were coming on to the market, and these were the precursors of the hollow-glass rods we see today.

TOP: A couple of nice Oriental Bay snapper caught in the early morning by Roger Craig from a rock just a kilometre from the middle of Wellington. *JIM O'BRIEN*

TOP LEFT: This 1962 photo shows members of the Wellington Surfcasting Club displaying a typical club day snapper catch. Most of these fish were caught surfcasting in Wellington Harbour. *JIM O'BRIEN*

LEFT: Frank Dellabarca and Jimmy Imlach, part of the Italian and Shetland Island fishing community at Island Bay, Wellington, unload a 1957 catch of groper, school sharks and snapper. *DOMINION POST COLLECTION, ALEXANDER TURNBULL LIBRARY, WELLINGTON, NZ. EP/1957/3049*

WELLINGTON HARBOUR

Pre-1960s snapper fishing was concentrated on the harbour for two reasons. Firstly, there were good numbers of fish, and secondly, access was easy by push–bike or tram; few people owned a car.

The favourite inner-harbour wharf was Pipitea. Anglers had to claim a spot early in the weekend, often arriving on Friday night with gear, food, drink and sleeping bag. They could be woken at three or four in the morning by late arrivals cursing that all the spots had been taken. At Pipitea, snapper catches were consistent, but nearly always between dawn and 7 am. Anglers used to time the end of the fishing by the 7 am ferry sailing past.

Very rarely were small snapper caught at Pipitea. Most were at least 3–4 kg and some 8–9 kg. It was possible to climb down the wharf to secure a fish, but it was a long way, so anglers carried gaffs made from several large hooks soldered together and attached to strong cord. Some days five or six big snapper would be landed, along with plenty of gurnard, kahawai and the odd kingfish. Jack mackerel could be caught by the sackful – and all this within a kilometre of Parliament Buildings. Unfortunately, Pipitea is now among those inner harbour wharves modified for containerisation and declared off-limits to the public.

Baits used for harbour snapper-fishing outside the hours of dawn and dusk had to be robust because of the legendary spotty populations. Fresh yellow-eyed mullet and good-sized spotty heads were favoured. When squid bait was available it became the number-one snapper bait. A lot of bait in the early days was salted, as freezers were uncommon.

Across the harbour from Pipitea Wharf lies Oriental Bay. This too was a prime snapper spot, especially fishing from the sea wall around to Point Jerningham, in shelter from the southerly wind. The fish here were smaller, and early morning was the most productive time. Snapper and plenty of other desirable species are still caught here, right in the heart of the city. Traditional deep-water possies are the reclamation at Greta Point, Evans Bay, Miramar Wharf, and rocky outcrops along the harbour's eastern side. Unfortunately Burnham and Shelly Bay, both renowned as good snapper spots, are now closed to the public.

The head of the bay at Cobham Drive is quite shallow, and before the airport extensions resulted in massive quantities of soft sediment being dumped into the bay, snapper-fishing was prolific. Early summer still has nights of good fishing along this stretch, but fish are typically 'schoolies' of around a kilogram. The deeper waters of Evans Bay have produced more snapper over the 20 lb benchmark than other harbour areas, and a recent 13 kg fish testifies that big snapper still travel the traditional pathways in Evans Bay.

PALLISER BAY

Nearly all modern-day snapper caught in an area between Corner Creek in the west, and a little past Whangamoana in the east, are typically in the 5–9 kg range. A photo from the late 1950s shows 23 good-sized snapper landed in just a

Glen Silverwood with a well-conditioned snapper of the magical 20 pounds caught from the Pukerua Bay headland in 1992. It was one of the last big snapper caught from this area, which had consistently produced the biggest shore-caught snapper in the Wellington region. *JIM O'BRIEN*

BIG SNAPPER TALES

While Wellington cannot lay claim to being a big-snapper region, some large fish have been caught. In the past, the Pukerua Bay headland consistently yielded snapper in excess of 20 lb for surfcasters and more of these benchmark fish were caught there than anywhere else around Wellington. An officially-weighed 14.52 kg NZ record snapper was caught here in the 1960s by W. Schonknecht. The same angler caught other fish close to the 30 lb mark from this area, but perhaps the biggest snapper to be caught in Wellington was caught by well-known veteran angler Lionel Renata. Lionel was new to the surfcasting scene, but had heard about a little beach just down from Boom Rock on the Makara coast where a few adventurous snapper anglers were catching some big fish. He made the one-and-a-half-hour tramp over the hilly country, finishing with a climb down the steep seaward cliffs carrying all his gear in a chaff-sack pikau. Lionel's gear was nothing flash: a three-metre solid fibreglass surf rod with a plastic surf reel, and a boat rod with a plastic centrepin reel. The line on the centrepin reel had to be stripped onto the beach and thrown out by hand.

Lionel fished until late but by midnight all he had caught was seaweed, so he decided to get some sleep. He baited up his boat rod with an outsized piece of paua and threw the line out into the channel. Wedging the rod between some rocks, he set a light drag on the reel, rolled out his blanket and went to sleep by the dying embers of his fire.

Waking around dawn, he saw his rod tip bending rhythmically, just like it had earlier when a hunk of seaweed on the end was being washed around by the tide. He went back to sleep for half an hour and woke to see the rod tip still bent, but this time saw a splash about 150 m out to sea. Frantically disentangling himself from his blankets, he covered the distance to his rod in bare feet, slipping over hard rocks. It took Lionel 25 minutes to work the fish back to shore through a jumble of rocks and weed. Its tail came out of the water close in, identifying it as a snapper, but he was not quite prepared for the massive fish that came ashore.

Like a lot of snapper anglers in those days, Lionel wasn't into weighing fish, but his catch bottomed out some scales which weighed up to 11 kg, with a good bit of the fish still on the beach. He has always regretted not having the weight of the fish accurately recorded or taking a photo. But when he carried the fish out, its tail was sticking out of the top of the pikau, and the weight of the fillets, boned and skinned, was 7.5 kg. This suggests Lionel's big fish was at least 15 kg.

few hours by members of the Wellington Surfcasting Club during an early spring fishing trip at Corner Creek. Even in those days anglers could still occasionally catch groper from the beach.

All along Ocean Beach snapper schooled, and multiple catches were not uncommon. Today it is more likely that a single fish will be caught, but that fish will almost certainly be over 5 kg. Snapper are caught along this coast in almost every month of the year, but the early spring and autumn months are the best. Crab is a good bait here.

MAKARA AND THE KAPITI COAST

The area of sandy, generally shallow-water beaches that run from Paekakariki north past Kapiti was very productive for early surfcasters. Certainly this area had its golden days, where families could drag for flounder by day and fish for snapper by night from the beaches of Paekakariki, Raumati, Waikanae and Te Horo.

In his diary, the late angler Jim O'More notes that in the 1959–60 season he caught 40 snapper, while his fishing cobber Sam Bratton caught 35. Not bad when you consider that fishing conditions on this coast are often marginal owing to the weather. The diary also records that Jim and Sam caught more than 300 snapper over six years.

ABOVE: A perfect end to a perfect day. Dave Gilroy with two fine 'pannie' snapper caught boat fishing at Pukerua Bay. Mana Island and a hazy Kapiti Island are in the background. Snapper of the size shown here are typical of most snapper being caught in the Wellington region as the stocks recover from the pair trawling of the1980s. JIM O'BRIEN

RIGHT: It is little wonder that with so many wide, open beaches, surfcasting and longlining with kites and electric Kon-Tikis are popular ways to fish along the Kapiti Coast. SAM MOSSMAN

Makara and Pukerua Bay boast a huge area of rocky and weedy shoreline interspersed with small stony and gravelly beaches. The waters here are ideal habitat for paua and sea eggs, prime food for big snapper. When Wellington surfcasters first started fishing this coastline in the mid-1960s, fishing for big snapper was very rewarding. There were few places that could not be fished by simply hoisting a pack on your back and either tramping round the coast or climbing overland.

Often the very big snapper were caught during the winter months, suggesting that the abundant food supply meant they did not need to migrate. Jim O'More was one of the first to fish seriously for snapper in the area during the late 1940s and 50s. He and a few keen mates would walk over the hills at the back of Johnsonville and camp at Pipinui Point, a short distance from the legendary Boom Rock. Here they could spear crayfish and collect paua without even getting their feet wet. The paua was used for snapper bait, and Jim's old photos show that they caught as many snapper as they could physically carry home in fillets – all caught on green cord handlines.

As mentioned, in the late 1970s the number of snapper caught by surfcasters in the Wellington region began to decline. A prominent industry representative stated that Wellington's snapper demise could be traced back to excessive pair-trawling by Nelson-based boats in the Pukerua Bay and Kapiti area. This is a common theme from many parts of the country, and many believe that commercial overfishing was the cause of the widespread decline. In the three decades since, the inshore fishery in the Wellington region has still not recovered.

FROM THE BOAT

Targeting snapper from boats is a relatively recent pursuit around Wellington, and grew with small-boat ownership in the region since the early 1980s. Anglers found that the key to successful snapper fishing was to use berley, and this revolutionised the sport. They still needed to find the spots and work out the best times of year, but attracting snapper and holding them at the boat was the secret.

Today, there is a thriving snapper fishery from Makara to the Kapiti coast, with the best fishing from February to May. Most fish caught weigh 1–2 kg, with a 3–5 kg specimen for every ten smaller ones. Often snapper on this coast are caught in water as shallow as 5 m, on the coastal reefs of the Kapiti coast or 'The Bridge', an area of foul between the mainland and Mana Island.

Wellington Harbour too has its keen snapper boat-fishers. Most snapper fishing in the harbour is carried out on the shallower eastern side where there seems to be more food available from a greater variety of habitats. The shallower water also allows for a slower rate of berley dispersal.

While the predominant fish size is again 1–2 kg, almost every year there is a report of a snapper over 10 kg being caught. A few seasons ago Mike Stempa got the shock of his life when, while fishing 2 kg line trying to catch a big elephant fish he reeled in a line-class world record snapper of 9.86 kg.

JULIE-ANNE CUNNINGHAM

THE TOP OF THE SOUTH

12

The 'Top of the South' as the Nelson, Golden Bay and Marlborough region is known, was an area favoured by Captain James Cook, who visited Ship Cove in Queen Charlotte Sound on all three of his exploratory voyages to the region. During his second visit in 1774, Cook was at anchor in his ship *Resolution* while his crew fished to supplement the ship's stores. Georg Forster, one of the naturalists on the expedition, described catching a snapper:

'Our sailors dragged a net [but were] more successful with the hook and line. Amongst others, a fine sea bream weighing eleven pounds [4.9 kg] was taken, it being one of those species which are to be met with in almost every ocean. It is caught on the coasts of England, in the Mediterranean, at the Cape of Good Hope and in the South Seas.'

It is fair that Foster should equate the New Zealand snapper to its similar-looking cousins in Europe and elsewhere, but the snapper of this region are a different species (see Chapter 1). They are the most southern breeding stock in the country. While snapper are also found further south on both the east and west coasts of the South Island, it is likely that they are mainly migrants rather than breeding stock.

A FRAGILE POPULATION

The Nelson/Marlborough region has big stable water masses in Tasman and Golden Bays, as well as the sheltered areas of the Marlborough Sounds, all suitable for spawning. Tasman Bay is the major snapper spawning area in the region, and also appears to be the southern boundary for breeding. The optimal water temperatures for good survival of snapper spawn may not be reached every year, or for any great length of time over the spawning season. When water does not reach these temperatures there is very little spawning success, resulting in erratic recruitment for this southern snapper population. Sometimes there are several seasons of poor spawning conditions in a row, making for a more fragile stock than those further north.

Snapper stocks in this area are the fastest growing in NZ. Sexual maturity of snapper depends on size rather than age, and occurs when the fish reach 22–25 cm in length. In Tasman Bay, sexual maturity can be reached in two years, compared to five years in the Hauraki Gulf. This quick growth and early maturity may be an adaptation to the unreliable spawning success caused by lower water temperatures, or perhaps the fast development of a greater body mass helps cope with lower water temperatures in winter.

The anthropologist Foss Leach has analysed Maori middens in the region, and tells a fascinating story in his book *Fishing in Pre-European New Zealand*. It is generally held that during the period of 1500–1800 AD, Europe experienced a cold period called the Little Ice Age, and there is evidence that this affected the southern hemisphere as well. Prior to this was the Medieval Warm Period, with temperatures slightly warmer than today. Sea temperatures here may have dropped by 2.6°C during this 300-year cold snap. If so, this would have greatly affected snapper spawning in Tasman Bay, possibly almost wiping out the species from the region for a time.

Aorere overlooking Wainui Inlet and Golden Bay, 1843, by an unknown artist. The numerous fish drying racks (whata) show the importance of fish to Maori of the time; in this case mostly barracouta, stingrays and a couple of snapper.

ALEXANDER TURNBULL LIBRARY C-030-019

Again, there is some evidence to support this. Fish bones in early midden sites dated in the warmer period before the Little Ice Age are 69–78% snapper, while later sites from the colder period have only 1–3%.

Snapper stocks in the Nelson/Marlborough region also show a lower genetic diversity than those further north, which maybe indicative of a 're-seeding' of the area by a small number of individuals. If so, these may have been fish with a genetic predisposition for fast growth – their bigger body mass coping better with the colder water, allowing them to be first to move back in and explaining the unusual traits of these stocks.

Another possibility is that a few snapper with fast-growth genes were best suited to survive the cold period in the region, giving them the most success at reproducing when sea temperatures rose again.

Sadly for this region, what the Little Ice Age may have done to snapper in pre-European times, commercial fishing came close to emulating in the late 1970s.

A SECOND DESTRUCTION

After World War II the gradual development of trawling saw the growth of a valuable year-round commercial snapper fishery in Tasman and Golden Bays. The small single trawlers tended to catch only the younger, highly variable age-classes of snapper. However, in the early 1970s, the introduction of pair trawling – towing a single large net between two boats – led to increased catches of larger snapper, especially during the summer spawning season. Pair trawlers are more effective at herding snapper into the net than single trawlers, and thus more efficient at catching larger fish.

In 1977, an Italian immigrant, Piero Rocco, who had married a young Marlborough woman, came to New Zealand. He had a fisheries research background, and took up a consulting position with the fishing company Sealords, based in Nelson.

One day, on a flight to locate surface schools of kahawai and mackerel, he saw some red smudges in Tasman Bay, off Motueka. He pointed them out to the pilot, who replied that they were krill. Rocco shrugged, unconvinced. The plane flew over Farewell Spit, then D'Urville Island and back over the red patches in the water.

Two weeks later, during trials with a new pair-trawl net, Rocco suggested towing the net through one of the reddish smudges. The result of just one tow was 43 tonnes of snapper. To some this would be a great catch, but Rocco was horrified: 'I was ashamed at what I had unintentionally done – those were spawning fish, congregations of breeding snapper.'

Word spread among trawlermen. The next day another pair of trawlers swooped and took 70 tonnes, and the following day another massive catch was taken. In just a few days, 230 tonnes were caught, and tragically a great many of the fish ended up being dumped at the Nelson rubbish tip because the factories couldn't process them all.

But a startling fact was the average size of the fish: a high proportion of them were much larger than caught by conventional trawl, the likes of which were normally only taken by line. It became clear that the boats were now

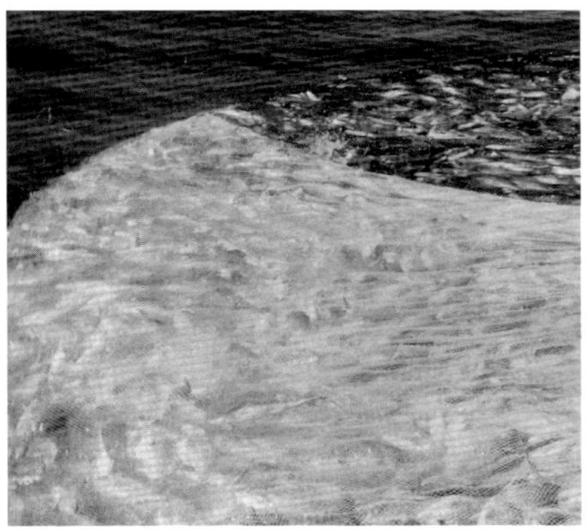

LEFT: Snapper were an early favourite with European settlers in the Nelson region. These three gentlemen were photographed with their catch of snapper at Cable Bay, near Nelson, around 1900.
ALEXANDER TURNBULL LIBRARY F-020648-1/2

BELOW LEFT: The first attempt at pair trawling for snapper in Tasman Bay in 1977 netted 43 tonnes of spawning snapper in one shot. Within a few years the resource had been almost wiped out.
PIERO ROCCO

BOTTOM: The excesses of pair trawling for spawning snapper in the summer of 1977–78 overwhelmed the on-shore processing facilities at times, and truckloads of snapper were buried at the Nelson tip, a tragic waste of breeding fish.
NELSON PROVINCIAL MUSEUM, G.C. WOOD COLLECTION 4225 FR23

fishing the large old fish that had previously been able to escape the old single trawls, and even the newer pair trawls. These fish were migrating into Tasman Bay to spawn in midwater, where they had in the past been out of the reach of trawlers.

The total catch that season for SNA7 was 2,720 tonnes, over a thousand tonnes more than had ever been taken before for the zone. In a written report in May 1978, Rocco protested at the folly and short-sightedness, predicting that at this rate of plundering the stock would collapse. Later in the year, a number of fishermen – in fact the ones who were making the largest catches – became concerned about the fishery. They approached the local MAF office, whose staff met with them and proposed new management controls. These met with a mixed reaction from a divided industry. The voice of optimism triumphed over caution and the quota for the most sensitive area, southern Tasman Bay, was set at 1,100 tonnes.

This proved to be far too high, as fish were concentrated in there from a much wider area. By 1980 the year's take for the larger region had crashed to 732 tonnes. By 1981–2 the stocks of all trawl-caught fish in Tasman Bay had been dramatically reduced, and recovery was very slow. Between 1990 and 2005 the commercial industry has only once been able to catch its greatly reduced quota, with catches in most years well under 200 tonnes.

Even the smaller spawning areas, in Kenepuru Sound and Golden Bay, were not immune. Kenepuru was intensively fished for several years by a group of fishermen using beach seine nets.

By the mid-80s, industry and the Ministry increasingly worked together to put in place management controls – closing nursery areas and vulnerable habitats – which, coupled with the catch restrictions of the QMS, are seeing a gradual but definite improvement of the Tasman Bay snapper fishery.

THE OLD DAYS

Old fishermen tell of seeing snapper in the surf at Rarangi near Blenheim in the 1950s, with the fishes' tails visible as they dug for shellfish. Shallow-water snapper fishing in Pelorus Sound was superb in the early 1970s. Barry Pomeroy of Golden Bay recalled fishing with handlines off the beach at Collingwood as a teenager, throwing the bait no more than 40 m out and catching several snapper averaging 5 kg.

'At times the snapper tails were waving out of the water, as they dug for cockles along Farewell Spit,' recalled Barry. 'We'd catch 40 or 50 snapper a trip.'

But Barry noticed that during the late 1970s the decline set in, coinciding with Piero Rocco's recollection of the plundering of the spawning fish in Tasman Bay. By mid-1985, the snapper fishery was at an all-time low.

But today there are some signs of recovery. Rocco says there was a very good spawning season in the summer of 1999–2000, when high water temperatures resulted in a good survival rate for eggs and larvae. The result was improved stocks of fish from an aveage of about 1.5 kg to 2 kg in 2007.

CLOCKWISE FROM TOP LEFT:
Mussel farms in the Marlborough Sounds are a great attraction for snapper, especially when the waste from harvesting is going back into the water.
SAM MOSSMAN

During the warm months some snapper make their way into Nelson Harbour, where a fishing jetty allows the public to fish in comfort along the main waterfront.
TONY ORMAN

Rob Jones with a fine Pelorus Sound snapper from among the mussel buoys.
TONY ORMAN

Piero Rocco, now a staunch conservationist, enjoys some 'R&R' in the Marlborough Sounds. *TONY ORMAN*

A few years ago MAF reduced the bag limit for recreational fishing to three snapper a day. Ironically, there was no matching reduction in the commercial quota – a point that rankles with local anglers.

COASTAL FISH

Despite this history of over-exploitation, anglers can still have some success in Golden Bay, although with reduced snapper numbers shore fishing is limited. The days of catching several snapper with handlines off Collingwood Beach have gone. However, a few favoured spots can offer reasonable fishing, such as Patons Rock reef, off the rocks by the mussel farm at Wainui Bay, and the Tarakohe breakwater.

In the Motueka area, Kina Beach, the north face of Rabbit Island, and the Back Beach at Tahunanui are spots for surfcasting. With a boat, more access opens up. The Boulder Bank near Nelson, Croisilles and D'Urville Island offer a multitude of spots.

A few snapper stray south of Marlborough , and some are taken each year in Cloudy Bay and Port Underwood. A key factor is the cool current pushing up along the east coast of Marlborough: if it's strong, snapper will not stray far south from Tory Channel; if it's weaker, Cloudy Bay and Port Underwood may begin to turn up snapper.

SNAPPER IN THE SOUNDS

The sheltered bays and inlets of the Marlborough Sounds are protected from the swells of the open sea, and even in gusty winds anglers can fish in the lee of a hill. Baches and holiday accommodation are dotted through the Sounds, Motueka and Golden Bay areas, and snapper charters based at Havelock and Nelson cater for the visiting angler.

Snapper are an enigmatic fish. Fishing for them, particularly in shallow water, is about finesse and focus. Darrin 'Woody' Woods of Blenheim is a keen snapper angler – when he gets time. 'Snapper fishing is not aggressive fishing. They're an instinctively wary and cunning fish. Technique and tackle are important factors in success.'

So too are patience and thought. Local knowledge is particularly important in the Marlborough Sounds. Keeping records and noting tide phases, time of success, moon phase and other relevant detail is invaluable and eventually patterns start to emerge.

It is important to keep an open mind. For instance, many believe that snapper move out of the Marlborough Sounds in winter, but Woody thinks winter is a great time to fish. The still, sunny winter days in Marlborough give optimum boating conditions. The only proviso is the winter fishing pattern changes the focus from deeper water to the shallower, warmer water.

Anglers need to adapt to the phase and strength of tides. When fishing deep water (30–40 m) a 6 oz ball sinker may be needed to reach the bottom. In shallower water, a 2 oz sinker is adequate. Woody uses a level-wind reel, 6–8 kg line, a one-metre trace of 18 kg nylon and a 6/0 hook, and a 2–2.5 m rod that gives a sensitive contact with the

Coral growing in 30 m of water off Separation Point in Tasman Bay provides an excellent nursery habitat for young snapper and tarakihi.
MIKE BRADSTOCK

SNAPPER SANCTUARY

When regular trawling began in Tasman Bay in 1946, fishermen quickly learnec to avoid certain places offshore where large lumps of 'coral' snagged their nets. These corals were not the same kinds seen on tropical reefs, but similar organisms called bryozoans or lace corals. A particularly large patch lying between Tasman and Golden Bays, known as the Separation Point coral bed, was not only a tricky place to fish around but also appeared to be a nursery ground, with huge numbers of small snapper and tarakihi. Clearly the coral was ideal habitat, with plenty of fooc, and nooks and crannies for small fish to shelter from predators.

For many years this area was unfished. However, by the 1960s fishermen had learned to make heavier trawl nets with chains, sledges and rolling bobbins that would ride over rough ground on the sea floor. It now became possible to fish over the coral beds, but the nets were very destructive to marine life on the sea bed. They undoubtedly contributed to the decline of snapper and other fish in the region as the coral beds were gradually smashed down.

After a study by Nelson fisheries inspector Frank Saxton in 1980 revealed the extent of the problem, MAF declared the big coral bed off-limits to trawling and dredging. Gradually tʰe coral grew back and the numbers of small fish increased. This was the first time in the world that bryozoans had been conserved to help protect a fishery.

baited hook. 'This enables you to sense takes on days when snapper are taking gently. Some days they take softly, pick the bait up and run towards you. Without that sensitive gear, you'd never detect it.'

Fresh, frozen pilchard is the best bait, with barracouta or fresh kahawai as backups. A berley bag is attached to the anchor rope at a practical depth so the berley is up-current and setting up a trail to where the baits are, out the back of the boat.

'If wind and current are in opposite directions, your bait may not be in the berley trail. Similarly in deeper water, having the berley at the surface means the trail is well above the bottom and therefore not effective.'

It is generally agreed that snapper move out of the Marlborough Sounds in winter, but a fair number are also resident during the cooler months. The late Rob Jones experienced good winter fishing action with Pelorus Sound fish in about 12 m of water on bleak, cold winter days with snow falling on the hills. Snapper have also been seen in Tennyson Inlet cruising a sandy bay on a still, winter-crisp mid-August day.

Myths abound in fishing. One story is that snapper fishing is confined to Pelorus Sound and a branch, Kenepuru Sound. One local who scorns that theory is Picton's Martin Tennant, owner of the Picton Sports Store: 'We've always caught snapper here in Queen Charlotte, even in midwinter,' he says, recalling a winter day when he and some friends fished for blue cod at Cape Jackson, then came back inside and had lunch by a mussel farm. While they ate, a snapper rig was cast out and caught seven good snapper. Like Woody, he believes there are some resident snapper all year, but undeniably the peak time is November to the end of March.

Martin believes in using plenty of berley and likes to pre-bait an area with feed over several days to get the snapper into the habit of feeding there. Then after six or seven days he will strike.

Mussel farms in the sheltered water of the Sounds are an attraction for snapper, and the fishing is especially good during harvesting, as noted in Chapter 7.

The new soft-plastic baits have been successful too. Martin and his son Travis score good snapper with designs like 'Nuclear Chicken', 'Pink Shine' and 'Lime Tiger' jerk shads.

Timing is a big factor in success, with dusk being a prime time for bigger fish. And if you want really big snapper, stay till after dark.

TOP: Some kayak anglers have developed sophisticated rigs. Stephen Tapp's kayak features rod holders, livebait tank, sounder, GPS and VHF radio. *STEPHEN TAPP*

ABOVE: The 'stealth' aspect of kayaks allows anglers to fish shallow water without spooking snapper. 'Shamus' boats a nice fish. *STEPHEN TAPP*

SUMMER HOLIDAYS

The Christmas/New Year holiday is a popular time for visitors to the top of the South Island. However, the abundance of people usually means a decline in successful snapper fishing. Snapper, particularly in shallow water, hate boats, jet-skis and water skiers fizzing around. They take to deep water or tuck up in quiet cul-de-sacs out of all the sea-traffic. Anglers should seek bays that have no houses or baches, and fish during quiet times like dawn or dusk when boating traffic is minimal. Be flexible and adapt to the prevailing conditions: knowledge of the vagaries of the fish and the area

Large tame snapper swarming around William Kerr, 1952. Although protected by law, these fish eventually disappeared and have not been seen at Double Cove for many years now.

ARCHIVES NEW ZEALAND/TE RUA MAHARA O TE KAWANATANGA WELLINGTON OFFICE AAQT 6401, A26609

THE TAME SNAPPER OF DOUBLE COVE

Locals in the Marlborough region are aware of the finfish reserve at Double Cove in Queen Charlotte Sound, and although it is not widely known elsewhere or even always acknowledged by marine reserve advocates, it may well be the oldest Marine Protected Area (MPA) in the country.

Researching its origins has proven difficult, but fragments of information suggest it that it began around 1948 when William Kerr began to feed the fish in the area. Eventually the blue cod and snapper became tame, and Double Cove became a port of call for tourist boats. This picture, taken around 1952, shows large snapper that could be fed by hand. Some even allowed themselves to be briefly lifted from the water without struggling.

The tourist attraction was eventually protected by statute, and the ban on taking finfish from Double Cove has remained to this day. However, the snapper have not been so fortunate, and it is presumed that they moved out to breed with others of their kind and were caught in the massive over-fishing of spawning snapper that occurred in the late 1970s. Tame snapper have not been seen at Double Cove for many years now.

will pay off. Some spots will also fish better than others depending on the tide.

A peaceful, relaxing form of snapper fishing is on shallow mudflats during dawn and dusk. The summer twilight is one of the better times, especially when coinciding with an incoming tide.

STEALTH-FISHING WITH KAYAKS

Snapper love cruising the mudflats of the Sounds and the shallow waters of Golden and Tasman Bay, which are rich in feed such as crabs and pipis,but stealth is required to catch them.

A kayak is just another kind of boat, but for a number of reasons kayak fishing has expanded rapidly in the last few years. Initially, kayaks were often used to gain access to offshore points, where they were easily pulled up on to the rocks: the actual fishing was done from the shore.

Around 2004 there was a rapid development of the use of kayaks as fishing boats in their own right, paralleling the rising popularity of sea-kayaking in general. An initial attraction is the low cost compared to a larger boat. Furthermore, they can be easily transported on a roof rack and dragged down a beach to launch where larger craft would not be usable.

Most kayaks on the market are made from rotationally-moulded plastic, which is much quieter in the water than aluminium, making it ideal for fishing shallow inshore waters without disturbing the fish.

As kayak fishing has become more popular, increasingly sophisticated fit-outs have become available on these small craft, including sounders, GPS, rod holders and VHF radio. However, some kayak fishermen prefer to keep things as simple as possible, perhaps revelling in the personal challenge of catching a good fish from a small and simple craft.

Jack McKenzie (uncle of Flag McKenzie – see p. 158) with a very serious snapper caught in the Waimea Estuary by his home at Monaco, Nelson, in 'the old days'. Courtesy of Flag McKenzie

JULIE-ANNE CUNNINGHAM

THE SOUTH ISLAND WEST COAST

13

A big question is just how far south do snapper go? The breeding population in the Nelson/Marlborough fishery is reported to spawn as far southeast as Cloudy Bay, while the odd fish has even been found in commercial catches around the Chatham Islands and even south of Stewart Island. On the west coast, fish are known to spawn in the Karamea Bight south to Westport. Stragglers have been caught at Jacksons Bay and even in Fiordland. Rarely, fisheries inspectors in Invercargill have been shown a thoroughly lost fish, but these are the exceptions to the rule: snapper like it temperate.

WILD WEST?

Like the north-west of New Zealand, the remote west coast of the South Island is one of the country's warmer and sunnier places. Despite its reputation for being wet, the Coast tends to rain heavily but briefly, and the clouds quickly peel back to reveal a blue sky once again. The sun-warmed surface waters of the Tasman Sea blown towards the land by the prevailing westerly winds create favourable conditions for snapper. The same fish that frequent the Sounds, Tasman and Golden Bays, and the west coast of the North Island, also wander down the West Coast.

Karamea Bight seems to be the top spot. Extending from Kohaihai past Karamea, Waimangaroa and Granity, to Westport, the Bight's inshore grounds are comparatively sheltered, stable and sandy – the kind of habitat snapper like – but further south than Cape Foulwind (41° 45') the coast is rougher for some distance and the numbers drop away. No doubt about it: the steep dumper-surf beaches, which are shingley and unstable, coincide with less-favoured snapper habitat. This is one place where the appearance of the shoreline seems to provide a good indication as to whether snapper will be found close by.

THE END OF THE LINE

There aren't any fisheries inspectors permanently stationed on the West Coast any more, so MoF isn't the first port of call for finding out about snapper. But keen surfcaster and Hokitika Fish and Game ranger Chris Tonkin is one of the many Coasters who would agree that the fishery is predominantly in the Buller district rather than Westland.

'Snapper become pretty scarce south of Hokitika – I surfcast here a lot but haven't caught one yet. Maybe it's because the rig and elephant fish get in the road, as snapper by comparison are pretty subtle feeders so perhaps they find the gear we use too heavy and won't bite at it so freely.'

Lee Kelly of Wildcat Charters fishes out of Greymouth and says he's only had three or four in the past seven years, 'but they were all big ones, 18 pounds or thereabouts. The main inshore fish we get around here are gurnard, rig, cod and kahawai.'

On the Coast, snapper are only seasonally close inshore. Most surfcasters agree they arrive about November, but that's probably an inshore migration rather than moving from another part of the country.

ABOVE: Having your face on a stamp is a sure sign that you have become a Kiwi icon. This snapper stamp was designed by Rick Youmans and issued by New Zealand Post in 1993. *COPYRIGHT NZ POST (1993), REPRODUCED WITH PERMISSION*

LEFT: Magnificent coastal scenery on the isolated northwest coast of the South Island. *SAM MOSSMAN*

BOTTOM LEFT: Hauling in – Steve Climo bringing in the goods at Mokihinui. *MIKE BRADSTOCK*

BELOW RIGHT: Fresh from the Tasman: three-pounders from Basil Climo's battery-powered Kon-Tiki longline. *MIKE BRADSTOCK*

'They're thickest in January and February, tapering off in March,' says Garry Anderson, who fishes all over the Karamea Bight. Mostly he reckons they're on the sand rather than rocky ground, eating pipis, crabs and the odd small fish.

'I got a 22-pounder last Saturday,' he says, 'Of course snapper are just one of the great options here, as well as big crays, 'puka, tuna and marlin.'

With no all-weather ports or sheltered harbours, both recreational and commercial fishing on the Coast tend to be strongly limited by weather. Combined with the low residential population, it's a saving grace for the fish. Garry has seen marks on his echo sounder that were so dense he thought he was over a rocky reef, but soon thought better of it as his lines brought up snapper after snapper and the 'reef'changed shape: 'It was just a solid mass of snapper.'

COMMERCIAL CONCERN

One thing everyone seems to agree is that the fishing has improved after the hammering it got from pair trawling in the 1970s and 1980s. There are always rumours like the story about a Nelson trawler that is supposed to have thrown away three tonnes of snapper off Westport in 2007, presumably because the owners had no snapper quota and didn't want to pay the deemed-value penalty. But it's impossible to know how true that story is, or how frequent incidents of this type are. Clearly though, Coasters particularly resent big trawlers from outside the region coming into 'their' coastal waters. In fact ever since the fishery management areas were set up in 1981 it's been a bone of contention that the West Coast was lumped in with Nelson and Marlborough. 'It's just not a natural division of the fishery limits,' says Flag McKenzie, who's been commercially fishing on the Coast for over 30 years, and surfcasting for longer.

Flag (short for Flagpole, a nickname he's had all his life on account of his lean and sinewy build) says he's always been able to find schools of snapper off Karamea during the winter months. In the 1980s he learned to recognise their marks on his echo sounder in 40 metres of water, and with a relatively short set net he could haul out 20 bins at a time. Being an environmentally-minded person long before he became a commercial fisherman, Flag resisted the temptation to flog a good thing too hard, and never fished the snapper when there was anyone else about. Besides, he knew these fish were among the few lucky ones that had escaped the pair trawlers, and he didn't like to see them pursued all year round, wherever they went. To maximise the catch value he made short sets so he'd always be lifting the gear while the fish were alive and kicking. This meant a top-quality catch that made export grade, and reduced the waste since dead snapper in set nets would quickly be damaged or completely hollowed out by sea lice.

SEEN ANY SNAPPER?

Waitangi Day on the West Coast. The sand is burning hot and the air is full of flax scent and salt. Cicadas scream incessantly from the nikau palms scattered on the shore terraces and rata trees that sprawl, mistily red-flowered, up the steep green hillsides. At Greymouth tiphead, anonymous shapes in the white water are surfers. Hectors dolphins pop up and steadily rotate in the foamy tidelines. Fish must abound here … but north and south there are miles upon miles of beaches and rock platforms with scarcely a surfcaster in sight. At the fish processor Westfleet, an inquiry about snapper is met with puzzlement. They haven't seen any for years. A lady in a fish-and-chip shop laughs at the idea of snapper and chips. 'Just elephant fish, love.'

At Westport there's plenty of tuna and trawl fish being unloaded, but no snapper. Out at the river mouth a couple of optimists are spinfishing for sea-run trout. Amazingly, even here there's not a surfcasting rod in sight. Given that the commercial snapper catch is negligible since the demise of pair trawling, could this be one place in New Zealand where they could be described as 'underfished'?

FAMILY FISHING

At Mokihinui towards dusk, Basil Climo and his family hit the beach with their torpedo-powered longline. It's a real family outing with wife Joan, son Steve and his family, plus a few miscellaneous friends joining in. Carl Fry would normally be skippering a toothfish longline boat in the Antarctic, but is taking time off; his partner Carol, who spent ten years on the factory trawlers, is now baiting the longline's 25 hooks – that's the maximum allowed for recreational fishing.

The 'torpedo' is aptly named: it slices through the waves as though they weren't even there, pulling the line out at a speed that would make any Kon-Tiki look feeble by comparison. It's just the ticket for fishing on a wild beach like Mokihinui, where line after line of breakers sweep in towards the shore and nothing with a sail would stay upright for two minutes. Basil's pretty optimistic about the prospects: the time and tide are right, and he's picked a likely spot where the wave patten offshore indicates there are plenty of holes where snapper like to feed.

Waiting for the fish to bite, Steve takes the opportunity to collect a ute-load of firewood, all in handy lengths that won't require cutting to fit his fireplace. Later he will lay his scaled and gutted fish on top of the load to drain off during the short drive back home.

Two sets of the torpedo longline yield eleven of what they call 'pansize' snapper – which means a fillet (not the whole fish) will fill a frying pan. These are sleek, gleaming three-pounders, full of fight and 'fresh as'. While the sun sets on the tiers of waves driven in by the now-rising tide, the team beat a retreat to the Climo house for a cuppa. The catch is shared around; nothing's wasted: there's no rough filleting, and someone's mum wants the heads. Joan tells a story about recently seeing some fish frames dumped in the creek and this draws mutters of disapproval at the waste as well as the mess.

We tend to think of snapper as fish of the north … but the fishery in the southwest looks healthy, even flourishing. It seems that a nostalgic photo taken of a person with a snapper under nikau palms might just as well have been taken in the South Island.

RICHARD ROBINSON

SNAPPER IN AUSTRALIA

14

Snapper are a greatly revered species in the central and southern coastal regions of Australia. 'Snapper have never been better in my lifetime,' says Australian author and snapper expert Steve Cooper.

Since 2000, anglers in Port Phillip Bay, Victoria, have been experiencing their best snapper seasons on record, with the 2006 snapper run rated the best for 15 years. Moreover, the news keeps getting better, with fisheries researchers predicting more good seasons. Says Steve:

'Snapper fishing is an adventure; a challenge that brings rewards for effort and is particularly suited to the single-minded angler. Fishing has been my life-long passion and, as with all pursuits, there are turning points. Mine came when I was 14 years old and 'doing time' at high school. With a classmate I started to fish at a small sandspit lagoon on the northern end of Corio Bay. Better known as the Grammar School Lagoon, it was to become legendary for its winter snapper.

'I would ride my bike to the Lagoon to fish at every opportunity. The ride took about 40 minutes and I must have looked a sight: fishing rods tied to the cross bar, tackle bag strapped to a carrier on the back, and a sleeping bag as well for an all-night session. As for clothes, we carried only what we could wear; there wasn't room for anything else.

'Tackle was different. Our reels bore names like Magna-Flite, Capstan, Surfmaster Avoca, Pfleuger Sea King and Penn Squidder. Rods started out as bamboo but eventually were upgraded to solid fibreglass, and we used the old cane rods to extend the butts on our rods so we could cast further.

'I love the adrenalin rush when that big, red-flanked snapper appears on the end of my line from the depths, dorsal fins erect and opalescent blue spots glowing like luminous beads along its lateral line. I still remember when I landed my first fish over 9 kg (20 pounds). It was on the high tide, three mornings after the new moon on 11 June 1967. Another school chum gaffed the snapper that pulled the scales down to 20 pounds 4 ounces or 9.18 kg.

'Those were happy days. Today, despite many changes, snapper still give me a buzz. Tackle has evolved greatly and today's snapper angler fishes mostly from a boat and has a sounder and GPS unit on board as standard equipment. It's a long way from the old pushbike-and-cane-rod days.'

SNAPPER SPEAK

Snapper devotees in Australia have developed a language of their own. In Queensland and New South Wales, anglers refer to smaller snapper as 'squire'; in Victoria, small snapper are 'pinkies', while in South Australia they are 'ruggers', and bigger fish (up to about 4 kg) are known as 'cobs'. They are called snapper only when they mature, and big, old specimens (especially those with big humps on their heads) are sometimes referred to as 'old man snapper', 'cockney' and 'red bream'.

Snapper are found right around the southern Australian coast, up the west coast as far as the Montebello Islands, and at least as far as Rockhampton on the east coast. The fishing is diverse, and every snapper sub-species has its own peculiarities.

ABOVE LEFT: Norfolk Island, a Pacific snapper outpost where these fish overlap territory with tropical species like sweetlip (called 'trumpeter' locally). *SAM MOSSMAN*

ABOVE RIGHT: Looking like a creature from another planet, the cuttlefish is a favoured prey for snapper after they spawn and die in NSW waters. *DR IAIN ANDERSON, AUTHOR OF THE SURFACE OF THE SEA WWW.ESC.AUCKLAND.AC.NZ/PEOPLE/STAFF/IAND002*

BELOW LEFT: Lawrie Birdseye with a nice 11 kg snapper caught off Point Lowly in South Australia's Spencer Gulf. *COURTESY OF STEVE COOPER*

BELOW RIGHT: Whyalla charter boat operator Rod North with a pair of monster South Australian snapper that weighed 13 and 15.5 kg respectively. *COURTESY OF STEVE COOPER*

WESTERN AUSTRALIA

Despite its huge size, Western Australia has its best snapper fishing concentrated in two main areas: Cockburn Sound in the south and Shark Bay in the north.

During summer, offshore anglers encounter snapper in large areas of deep water. Perth anglers often catch fish up to 4 kg in water only 100 m and about 25 km offshore. However, the biggest specimens come from around shallow inshore reefs and structures.

Perth tackle identity Hal Harvey says the northern cut-off for snapper used to be Coral Bay, about 100 km south of Exmouth, but in recent years they have been caught as far north as Broome. He says there are at least three genetically separate breeding populations of snapper in Shark Bay: the Eastern Gulf, the Western Gulf, and oceanic stocks. The snapper that make up these stocks do not migrate: they have a limited home range in which they forage, and do not interbreed.

West Australian snapper form breeding aggregations, and WA Fisheries says this is when anglers tend to catch significant numbers of larger fish. In northern waters snapper breed from late May to August, while southern stocks breed between October and early March.

Lou Rummer fishes for snapper around Cockburn Sound, where fish enter through the narrow north entrance and work their way clockwise around the sound. At any one time there can be up to 200 tonnes of snapper there – a huge number of fish for an area only about 5 by 9 km.

Lou says the big attractions in the sound for snapper are mussels and crabs. Snapper are caught there most of the year, but the best time to target them is from September through to December when they enter in vast schools to breed. The area is closed to fishing from mid-September through to the end of October to protect these breeding fish. The fish caught range from the just-legal size of around 2 kg, through to monsters of 17 kg. The surrounding offshore reefs also carry plenty of snapper.

SOUTH AUSTRALIA

When Australian anglers talk big snapper, most think of places like South Australia and Spencer Gulf, where a 9 kg snapper is not rare and every year fish over 15 kg are caught. The South Australian snapper season runs from spring to autumn, and the hottest snapper-fishing months are November to March, according to Adelaide-based fishing identity Jim Harris. In the spring, snapper migrate into the gulfs and estuaries. The Gulf St Vincent and Spencer Gulf are where the bulk of the snapper fishing takes place, mainly over offshore reefs.

Snapper reach the state legal minimum length of 38 cm at about six years of age. Spawning takes place between late October and early March when the surface water temperature is about 18°C.

The waters of the Spencer Gulf (and further west at Streaky Bay) are renowned Australia-wide for regularly

AUSTRALIAN SNAPPER HOTSPOTS

yielding snapper over 10 kg. Pumping out berley and watching snapper feed on the surface is not something anglers in other parts of Australia expect.

Jim Harris says Point Lowly, north of Whyalla in Spencer Gulf, is truly a big fish location. He was in a party of four anglers who caught 14 big snapper of 9.8–15.6 kg in just three hours, using large cut baits of 'sea-salmon' (kahawai). Jim says the big snapper run in Spencer Gulf starts in November (when the snapper season opens) and can last through to April. However, there are always snapper there, regardless of season. The top of the Gulf from Port Hughes to Port Augusta holds good numbers of smaller snapper. Anglers believe that when they reach about 6 kg, snapper move out of the Gulf.

Spencer Gulf is littered with unauthorised artificial reefs. Car bodies and refrigerators are common, and there are tyre reefs at Wallaroo and Whyalla. At Port Broughton the local council put down a reef of car bodies, and the Gulf has a few divers' reefs, as well as a number of shipwrecks.

In Gulf St Vincent there is a sunken barge which is a popular hang-out for snapper, and large fish are often caught there in December and January. The outer harbour shipping channel off Adelaide produces some of the state's biggest snapper, up to 17 kg.

VICTORIA

Snapper are caught all along the Victorian coast, but the best results come from Port Phillip Bay and Western Port. Moreover, while most snapper are caught from boats there are a host of piers, jetties and breakwalls to satisfy land-based anglers.

In both bays, the migration begins in spring and the season runs through until March or April. The peak fishing months are October and November. The annual early-spring snapper run in Port Phillip Bay is among Australia's best-known fish migrations. The good news for anglers is that in this area the movement of snapper schools is more obvious than elsewhere, so catching more and larger fish is easier. Even though snapper over 9 kg are caught every year, most are 1.5–4 kg. The reality of snapper fishing here is that those who wish to target fish of 5 kg or more need to head for water at least 18 m deep. They won't catch so many, but the quality is much better.

The snapper migration begins as a steady trickle rather than a red flood. These fish follow the same set routes as they filter through Port Phillip Heads, working their way through the gutters between the inshore reefs. Once inside Port Phillip Bay the schools appear to split into two different groups; the larger group moves anti-clockwise along the eastern shore while the smaller group runs along the western shore.

Western Port is smaller and quite different. Here the bay has a tidal flow that can push water through its eastern entrance at up to 7 knots; even with 280 g (10 oz) leads anglers have no hope of hitting bottom here when the tide is moving.

LEFT: The late Ted Semple, one of the owners of the fish factory at Norfolk Island, is pictured here with a catch of snapper he made, hanging from the front of his Honda truck. The largest of these fish weighed over 23 kg. *COURTESY OF IAN KENNY*

BELOW CENTRE: Hyperostoses or exostoses (bony overgrowths) are the rounded bone expansions often seen on the lower vertebral spines of snapper from both sides of the Tasman. *SAM MOSSMAN*

BOTTOM LEFT AND RIGHT: Smaller cousins of the snapper in Australia include the pikey bream and yellow-fin bream. A single specimen of the yellow-fin bream is recorded from New Zealand waters. *SAM MOSSMAN*

The best snapper fishing is in November and December, although fish are caught from late September through to April. Most anglers time their trips to coincide with the tide when water flow is reduced or stopped.

In recent years, the hottest fishing has been off St Kilda, Carrum, Frankston, Seaford and Mornington. When the snapper are running here they are also often caught on the opposite side of the Bay, at St Leonards, Corio Bay and Werribee.

Victorian snapper are split into eastern and western stocks, with different migration and spawning patterns. The two groups separate at Wilsons Promontory: the western fish migrate west along the coast towards South Australia, while the snapper in eastern Victoria mix with those in New South Wales. It is thought this separation was established during the last ice age 10,000 years ago, when lower sea levels created a land-bridge between Victoria and Tasmania, isolating these fish. If this is so, they don't seem to have got over it yet. The cooler waters around Tasmania mean that snapper are not common there, with only a few being encountered in the warmer waters of the north-eastern corner.

NEW SOUTH WALES

New South Wales offers many options for catching snapper. Perhaps the biggest change in recent years, as in New Zealand, has been the advent of soft-plastic lures and the way offshore anglers have been using them.

On the far south coast, from the Victorian border to Bermagui, the Christmas period is good for snapper fishing. Most fish along this part of the coast are to about 3 kg, with the odd fish making it to double figures. The reefs off Merimbula, Eden, Green Cape and Disaster Bay can be fruitful. Many of the rocky ledges, particularly where there is a big wash, will produce fish of 2–4 kg. At Bermagui, Four Mile and Six Mile Reef sometimes produce good bags.

Moving up the coast, the Sir John Young Banks, about 11 km east of Greenwell Point, is highly regarded for its snapper fishing, as is the Shallows, about 7 km east of Currarong.

Wollongong-based angler Greg Clarke fishes for snapper from June to August over the inshore reefs to the north and east of the Bellambi boat ramp, taking advantage of the breeding cycle of the cuttlefish that move onto these reefs during winter to spawn and then die. Spent cuttlefish carcasses float to the surface, attracting snapper to the surface to feed. Greg's technique relies on stealth. He motors along the shore from the back of the breakers, out to about 15 m of water, looking for tell-tale signs of snapper feeding on the dead cuttlefish. 'When I see a cuttlefish body lift out of the water or move sideways I know there are snapper beneath it,' Greg says. He positions his boat downwind and casts unweighted cuttlefish as bait to the scene of the action. After casting, he waits a few minutes and if nothing happens, repeats the process. He has also had success using large white single-tail soft plastic grubs or white Deceiver flies delivered with an intermediate-density line.

The reefs east and north of Wollongong produce plenty of snapper, some as large as 9 kg. The best method is to anchor and run a berley trail of unweighted pilchard or cuttlefish.

Near Sydney, May to August is best, while autumn to spring is regarded as best from the south of Sydney to Shoalhaven area. In the Sydney area the biggest fish are offshore, according to local angler Gary Brown: 'The snapper move in around mid-May and stay until late August, in conjunction with the cuttlefish run. They turn up a month or so earlier at Narooma for the same reason.' Most of the big fish (around 9 kg) are caught on the offshore reefs. Snapper of 2–4 kg are considered a good size at Port Hacking, Botany Bay, and the Sydney Harbour entrance.

'Snapper move inshore close to the rocks after we have had a big sea running,' Gary says, 'and land-based anglers fishing spots like the rocks on the Kurnell Peninsula can do well.'

After August, the westerly winds knock the seas flat and the small snapper start moving into the estuary systems. At Broken Bay, Barrenjoey Headland on the south side, and Box Head on the north can fish well. When the water cools in May, snapper from 2–4 kg are caught during the day.

Good spots in Sydney Harbour include Quarantine Bay, Sow and Pigs Reef, The Wedding Cakes, South Head and North Head. Most of these snapper are up to about 2 kg.

Newcastle-based angler Mark Williams says that the more difficult the area is to access, the better the fishing is likely to be. Many serious snapper anglers prefer to fish Port Stephens because there are more snapper there.

QUEENSLAND

Queensland angler Jason Comino fishes what he describes as an underrated snapper fishery in Moreton Bay, near Brisbane. He says he has experienced some great sessions, including one that saw him catch five snapper between 6 kg and 10 kg. The big snapper show up in May at the northern end of the Bay and run until late August. 'But you can still get the occasional big one from September through to May,' Jason says, 'I've pulled fish to 10 kg in February; it's just that there are a lot of bigger fish during winter.' Further north along the coast, most snapper are caught on the offshore reefs along with other species such as pearl perch and red emperor. Snapper numbers thin out fast as the water warms to tropical temperatures.

NORFOLK SNAPPER

Smaller snapper are caught around Norfolk Island itself, indicating some breeding in the area. Norfolk waters may be considered a cross-over zone between temperate fish such as snapper, and tropical species such as sweetlip, known locally as 'trumpeter'. Sweetlip and snapper are in competition for the same feeding niche, and the sweetlip seem to have the best of it, being by far the more common catch of the two species in the waters around Norfolk.

Some very big snapper have been caught by Norfolk-based fishermen venturing up to 35 nm south of the island along a shallow extension of the Norfolk Ridge. Local fisherman Ian Kenny has caught snapper in this area weighing up to 42 lb, and other catches include a specimen reputed to weigh over 23 kg (50 lb). It is possible that these very large old fish have slowly worked their way north from New Zealand.

SNAPPER COUSINS

Other members of the Sparidae family found in Australia include bream, a smaller fish than snapper. There are three common types – black, yellow-fin and pikey bream – as well as a northern cousin called tarwhine. All are growing in popularity as recreational species, especially since the advent of lure fishing (particularly with soft plastics), and a professional tournament circuit has been established.

In August 1990 a single yellow-fin bream was caught in New Zealand, by a trawler in 50 m of water off Spirits Bay in Northland. It is thought that this fish may have been a hitchhiker in a ship's ballast water, as the eggs and larval stages of these fish last only four weeks before settling on the bottom – not long enough to cross the Tasman.

Queensland charter operator Matt Johnson supplied this image of what may be the world's ugliest snapper, caught by one of his clients from a reef off the Gold Coast. 'Knobby' snapper are not unusual in parts of Australia, but this is an extreme example. ***COURTESY OF MATT JOHNSON, WWW.FISHTHEDEEP.COM.AU***

GETTING THE HUMP

Some Australian snapper develop peculiar, exaggerated humps on their heads. Big hump-headed fish are most common off south-east Queensland, New South Wales, Victoria and West Australia. Many fish of equal size in South Australia have no hump. However, even in areas where hump-headed fish are common, not all fish have them. It seems to be mostly reef-dwelling fish that are humped.

In New Zealand some big old reef-dwelling 'moochers' develop thickened noses and modest humps on the head, but certainly nothing approaching the spectacular crests seen on some Australian fish. The thickened noses and lips may be scar tissue, but the hump is a distinct bony outgrowth of the skull. In parts of New Zealand these bone growths are known as 'tilly bones', and pre-European Maori sometimes carved them or used them as pendants.

Another abnormality commonly seen on snapper from both sides of the Tasman is the rounded bony lumps on the lower vertebral spines, called hyperostoses or exostoses.

It appears that little research has been done on these growths. In his book *Fish Might Fly*, Australian marine scientist Dr Julian Pepperell suggests they are a reaction to a parasite, bacteria or v rual infection associated with reefs. New Zealand fisheries scientist Dr Larry Paul notes that similar bone growths also occur in several other fish species, and may be calcium-storage sites.

SAM MOSSMAN

KAIMOANA

15

There are many reasons why snapper are an iconic fish in New Zealand. They are a common species, plentiful in the areas of highest human population. With a liking for coastal habitats they are easily accessible to anyone fishing from the shore or from small boats. They are handsome, fight well without being too difficult to catch, and can be caught by a wide variety of techniques. But ultimately, the snapper's position as 'the people's fish' rests mainly on its reputation as a table fish.

PRE-EUROPEAN COOKING AND PRESERVING FISH

In the north, fish were an important part of the diet for early Maori, probably accounting for about a third of their total meat intake. This fraction was lower in the south, where moa and marine mammals were more important. Compared to other fish commonly eaten by the early Maori, snapper contained high levels of protein and energy. Maori could not survive on fish alone, as it did not contain carbohydrate. In the north, kumara partly supplied this, with fern root more common to the south where colder temperatures prevented kumara cultivation. Energy could also be obtained from the fat of sea mammals, sharks livers, mutton birds, freshwater eels and (later) pigs.

After they made their catch it appears that ancient Maori usually scaled fish using tuatua shells. It seems they did not fillet their fish, but cooked and ate them whole, in the Polynesian fashion. This had the dual benefits of avoiding any waste, and accessing the bulk of the fish oils. Snapper has 5 g oil per 100 g, most of which is in the frame (head and backbone). Parts of the gut were also eaten, and even quite small snapper could provide a meal when eaten whole.

Fish were most often baked in a hangi (ground oven) in a manner still practiced today. Steaming is another method recorded: the fish was wrapped in leaves and tied with vines, then placed on hot stones and turned occasionally.

Fish were also grilled over an open fire when the people were in a hurry and could not be bothered with a more formal preparation. Fish were (sometimes) gutted, and then run through with a stick which could then be stuck in the ground near the fire until one side was cooked, then flipped over.

Huge amounts of fish were dried and preserved for winter, often at seasonal fishing camps. They were cleaned, split to the backbone and hung up to dry on a wooden frame called a whata. Salt was apparently not used to preserve fish in pre-European times; just the wind and sun. Dried fish was stored in baskets and sometimes traded to other tribes (often inland) for birds, workable stone and produce.

Seafood was an important source of iodine before salt was iodised in the late 1920s. Lack of iodine in the diet can result in goitre. An examination several years earlier of inland Tuhoi (Urewera) children found 30% thyroid enlargement, while coastal-dwelling children of the same iwi (with a strong seafood diet) were practically free of goitre.

PAKEHA AND SNAPPER

By contrast, Pakeha New Zealanders of a century ago were not great fish-eaters. In 1914 the per capita consumption of

For casual meals, Maori often grilled fish on a stick over an open fire, as seen in this scene at Pakanae on the southern shore of the Hokianga Harbour around 1828. The artist, Augustus Earle, and his friend Mr Shand are shown offering ribbons to the two women cooking. *ALEXANDER TURNBULL LIBRARY, WELLINGTON. PUBL-0015-04*

fish was only 2 kg per year, as against 20 kg today. Meat was cheap and widely available, and a poor inland transport network made it hard to get fresh fish to markets far from coastal ports. Without easy refrigeration, long-term storage relied on salting, drying or smoking.

Rather than a delicacy, fish was seen as a cheap, easily available source of protein. From a slender volume called *How to Cook New Zealand Fish*, published in the late 1920s, comes a comment on the attitude of Pakeha to fish at the time: 'In New Zealand, except at a few expensive hotels and boarding houses, very little variation is shown in the method of cooking fish. Frying and boiling are the only two methods in general use.'

Preserving fish in sealed jars was a popular way of storing it for long-term use before the era of home freezers, although snapper was not necessarily the most popular species for this. Even into the late 1950s '... there was little variety in food and, apart from a small number of Chinese restaurants in the main centres, nowhere to eat out other than hotels or 'greasies' – dining rooms that served fish and chips or steak and chips,' according to the historian Michael King.

But a relaxation of restaurant licensing laws in the 1970s, the large number of Baby Boomers reaching adulthood, and the availability of cheap air travel to the rest of the world in the 1970s brought new and liberal ideas – including ideas about food. A rash of formerly exotic foods became available in restaurants, and seafood became popular and even trendy as its health benefits were recognised.

In 1969 around 80% of the national catch was from 10 species. Snapper was number one at 35%. As a much wider range of species were exploited and aquaculture developed in the following decades, a much wider range of dishes became available. Forty years ago most New Zealanders would never have considered eating squid, octopus or raw tuna.

With the advent of the deep-sea fisheries snapper became less important as far as commercial exports were concerned, but the high value that domestic consumers placed on this species was maintained. The difference was that New Zealanders now had to pay export prices for their favourite fish in a booming international market. This drove the price of snapper well above that of beef or lamb, making it a luxury food – at least for those who were not able to catch their own. Fish and chips are still the New Zealand's most popular takeaways, but the price of snapper (over $30 per kg at the time of writing) now precludes its use in the fish-and-chip trade, and inferior bulk-caught species such as hoki and southern blue whiting are much more likely to be used.

Today the more expensive snapper is likely to be found in an up-market seafood restaurant, and may be cooked whole or used as the centrepiece of a complex dish. The seafood industry continues to promote increased consumption of seafood, stressing the health benefits of Omega 3 in fish.

LEFT: The hangi or ground oven is a traditional Maori method for cooking in quantity. Here men close a hangi with earth at Te Kaha in 1944.
JOHN PASCOE COLLECTION, ALEXANDER TURNBULL LIBRARY, WELLINGTON. F-001075-1/4

BELOW LEFT: Fish and chips have long been New Zealand's favourite takeaways, and rugby the favourite game. Here two spectators enjoy a meal before the British vs Maoris rugby match at Athletic Park, Wellington, in August 1950.
EVENING POST COLLECTION, ALEXANDER TURNBULL LIBRARY, WELLINGTON, NZ. 114/181/01

BELOW: Today, with snapper costing more than $30 per kg, it is seldom used in fish and chips unless specifically ordered.
SAM MOSSMAN

FILLET OF DOUBTLESS BAY

A good example of the use of snapper in haute cuisine is a dish developed by Scott Fraser, Executive Chef at the five-star Carrington Resort on Karikari Peninsula in Northland. Carrington is a quietly sophisticated resort with its own golf course, skeet range, stables and pool. It also has its own vineyards and a winery, as well as farming its own beef and honey.

Hailing originally from Scotland, Frazer has cooked in France, Spain, California and Australia before moving to the far north of New Zealand. He has trained under Michelin chefs and delights in creating dishes from fresh, high-quality local ingredients. What could be better suited to Northland than a snapper dish?

'The crushed peas-and-carrot foam in this dish is a modernised version of the peas and carrots that my mum used to serve me when I was a kid. The sweetness on the vegetables and melt-in-the-mouth pork cheeks seek to match the characteristics of the fantastic snapper – in Spain it is common to match fish and meat – all of which is countered perfectly by the sour and slightly bitter pink grapefruit. In the restaurant we match this dish with our 2006 Karikari Estate Chardonnay.'

Fillet of Doubtless Bay snapper with marjoram crushed peas, sticky pork cheeks, carrot foam and pink grapefruit powder (Serves 6)

Scott Fraser, Executive Chef at the five-star Carrington Resort on Karikari Peninsula in Northland, at work on the snapper dish. *SAM MOSSMAN*

6 x 130 g snapper pieces
6 pork cheeks, trimmed of all fat and sinew
1 onion, diced
1 small bunch of thyme
2 cloves of garlic
1 leek, diced
1 stick of celery, diced
1 litre chicken stock
1 bay leaf
500 g bag of frozen peas
30 g butter
50 g of marjoram (or oregano)
1 pink grapefruit
300 g carrot, diced
1 sheet of gold gelatine

The day before:

Using a potato peeler, peel the zest of the pink grapefruit into strips. Put onto a baking tray and leave in a warm place to dry out completely. If your oven has a pilot, this is a perfect place for it; at the restaurant we place it on top of the grill. Once completely dried out, blend in a coffee grinder until you get a fine powder.

To cook the pigs' cheeks:

Heat a heavy-bottomed saucepan on the stove until smoking and brown the pigs' cheeks on all sides. Add the celery, onion, leek, bay leaf, garlic, thyme and 100 g of the diced carrot. Sweat the vegetables until softened, then add the chicken stock. Bring slowly to the boil and remove any scum that rises to the surface. Cook on a very gentle heat until the meat is very tender (about three hours). Remove the cheeks and reserve. Pour the stock through a sieve into a clean saucepan and discard the vegetables. Bring the juices to the boil and reduce by about 4/5ths until the sauce has thickened and become sticky. Add the reserved pigs' cheeks and keep warm.

For the peas:

Bring a pot of salted water to a rolling boil and cook the peas until soft. Drain, then crush with the marjoram, butter, salt and pepper. Keep warm.

To make the carrot foam:

Cover the rest of the diced carrot with water and bring to the boil. Simmer until very tender, drain, then blend until completely smooth. If it does not blend well add a little of the cooking water. Season with salt and ground white pepper and keep warm.

Soak the gelatine leaf in cold water until soft, then dissolve in the warm carrot purée. Put the purée into an Isi cream whipper and charge with a nitrogen bulb. Chill in the fridge, shaking every 20 minutes until cold.

To finish the dish:

Grill, pan-fry or poach the snapper. Place the peas on warm plates and top with the pigs' cheeks. Put the snapper to one side and top with the carrot foam. Sprinkle the pink grapefruit powder over the plate. Serve immediately.

ABOVE: Fillet of Doubtless Bay snapper with marjoram crushed peas, sticky pork cheeks, carrot foam and pink grapefruit powder. A high-end snapper dish created by Scott Fraser, Executive Chef at the Carrington Resort. Today, snapper is as likely to be part of a gourmet dish as a simple fried fish meal. *SAM MOSSMAN*

BELOW FROM TOP TO BOTTOM:
Snapper roe wrapped in bacon and smoked is a delicacy. *SAM MOSSMAN*

After filleting there is still a lot of meat left on the frame. Separated at the joints, breadcrumbed, seasoned and cooked, this makes delicious finger-food at a barbecue.

Often overlooked, snapper 'wings' or 'throats' contain much delicious meat. This one has been split and hot smoked. *SAM MOSSMAN*

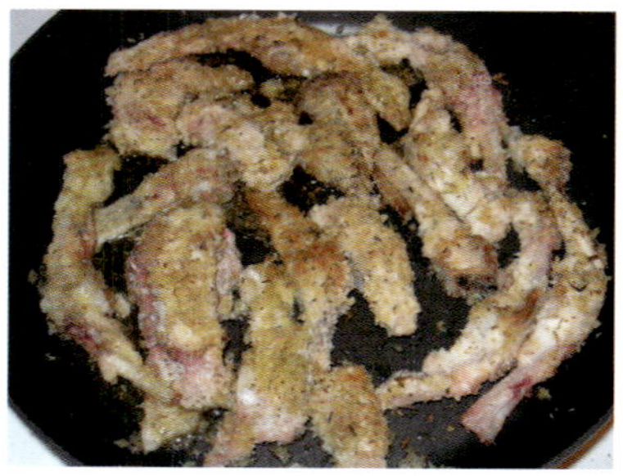

THE PEOPLE'S FISH

Despite the new-found celebrity status of seafood, the snapper still wears the mantle of a bread-and-butter species, at least in the north. Because the human population is higher in areas where snapper are common, there is no other single species of fish that receives so much attention from amateur fishermen.

Recreational fishing has experienced several booms in the last few decades and supports a multi-million-dollar industry that includes importers and retailers, boat and trailer builders, mechanics, travel, accommodation and bait suppliers, just to name a few. When this economic activity is added to the social benefits, relaxation and recreational enjoyment for a large part of the community it seems that recreational fishing is very important to New Zealand. Surveys have shown that 20–33% of people go fishing at least once every year.

With the high prices for snapper over the counter, catching your own has never been more popular. Not only does fishing offer the opportunity to wind down from modern-day stress, it gives our increasingly urban population a chance to get in touch with nature, with the potential bonus of putting a feed of world-class seafood on the table. With an increased perceived value of fish, recreational anglers are in general now taking much greater care of their catch and maximising its usage. A generation ago, fish were often stored in a wet sack or a fish box of some sort, with no refrigeration. These days it is unusual for a boat not to have an icebox of some kind to store the catch, and snapper are often 'ikied' – killed with a spike to the brain – to improve their eating quality before putting them on ice.

Grilling or baking whole snapper in the Asian fashion is becoming more popular now, and this method also makes good use of the whole fish. However, many people still prefer to fillet their fish. This can be a wasteful practice, but there is a lot more that can be done with a snapper to utilise it more fully.

DISASSEMBLING A SNAPPER

During the spring and early summer months when snapper are preparing to spawn, the females are full of roe. Smoked snapper roe is a great treat spread on a hot piece of toast with melting butter. When harvesting roe, try to avoid breaking the fragile membrane that contains the roe sac. This requires care, and it helps to be able to tell males from females when processing. The best way is to look at their throats: during spawning season the throats of male snapper exhibit rusty brown-grey to black markings, while those of the females are usually white or pale grey. This is a comparative thing, and most distinct on freshly-caught fish in spawning season.

Open the gut with a shallow cut along the belly, then cut upwards behind the pectoral fin (this is one of the first cuts normally made when filleting.) With care, the roe can be extracted intact by breaking the connections at the top and bottom.

Many people prefer boneless fillets – food without fear. A little care when filleting will result in greater flesh recovery. Cut well up into the back of the head and through the rib bones to take the belly-flap off with the fillet.

BELOW: Keeping fish in top condition starts from the moment they are caught. These snapper have been killed with an iki spike to the brain and stored in ice to chill them quickly. It really does make a big difference. *SAM MOSSMAN*

BOTTOM: When collecting snapper roes during the spring and early summer months, it is useful to be able to tell the male and female fish apart. During the spawning season males develop grey-brown colour under the head area, while females' throats remain white. *SAM MOSSMAN*

The knife can then be worked down just under the bones that surround the gut cavity, to save as much flesh as possible before skinning the fillet and removing the narrow centre strip to render a fillet boneless. Of course, it is not necessary to remove the skin: you can scale the fish first, and leave it on. Some people turn up their noses at fish skin, but others consider it the best part, and it certainly is particularly rich in Omega 3s. The trimmings from around the belly of a snapper are particularly rich, and ideal for making sashimi. Often you can simply slice thin strips off from around the rib bones and serve them in the traditional Japanese style with soy sauce, wasabi and grated ginger and citrus rind.

WASTE NOT, WANT NOT

Many anglers are not very careful about making optimal use of the fish they catch. They take fillets off and throw the rest away. There is a lot more mileage to be had out of such a noble fish.

Firstly there is a large amount of flesh between the head and the body of a snapper, known as the 'wings' or 'throat'. If you are cooking a fish whole, this is one part you should be sure to pick over well. The meat in the wings is the high-energy muscles that power the fish's pectoral fins, so it is denser, darker and tastier than the fillet, and the bones in it are large and easy to remove after cooking.

To remove the wings after filleting, hold the fish upside down and first cut through the throat at the narrowest point where it joins the head just behind the lower jaw. Then sever the membrane between the wings and the gills, pull hard backwards on the wings and twist so they come away from the head. You will get a roughly V-shaped portion with the pectoral fins attached. Clean this by scraping off the scales then cutting through the lower bones so it will lie flat. A favourite way of cooking wings is by hot smoking (see below), but they are also excellent fried or steamed.

A snapper head also contains a significant amount of meat and can be boiled or smoked. The easy parts to get at are at the top of the head and in the cheeks – finger-licking good! – and some people even eat the eyeballs.

There is also a surprising amount of flesh left on the frame after filleting so the frame of a reasonable-sized fish can be turned into another delicious finger-food. Sever the frame at the back of the skull, and at the tail. Cut off the dorsal and anal fins – shears or heavy scissors are good for this. Now sever each vertebra at the joint with a knife and cut away top and bottom, following the sweep of the bones. These pieces can be coated in breadcrumbs or herbs and spices, then thrown in a pan or on the barbecue. They only need about two minutes to cook, but at a barbecue it is often hard for the cook to keep up with the eaters. The flesh is delicious and can be just sucked off the bone.

The head and frame can also be used to make stock or soup, as can the left-over bony sections from the fillet, although this is often the share of the family cat. In some Asian countries fish skin is cooked in super-hot oil to produce a crunchy snack similar to potato crisps. Snapper liver is also a favourite with some.

ABOVE FROM TOP TO BOTTOM:
Smoked snapper is a Kiwi classic. Here split snapper drain after brining, prior to smoking. *SAM MOSSMAN*

The smokehouse loaded with snapper, gemfish and snapper roe, ready for a slow smoke. *SAM MOSSMAN*

After five hours of slow smoking, Ross Grieve is delighted with the final result. Now comes the best part ... *SAM MOSSMAN*

SMOKING

Snapper can be split and smoked too, and then all the flesh can be picked off the frame for maximum benefit. Salting and smoking fish was initially a way of preserving them in the days before refrigeration, but the smoky taste, the salt and other flavourings (brown sugar being the most common one these days) have a great appeal.

Internationally, many dishes that are considered regional classics were developed out of necessity. Recipes like haggis and gumbo were developed as a way of cooking the bits that were left after people higher up the pecking order had taken the best parts. In the same way, although smoking is seldom necessary to preserve the fish today, smoked fish is a delicacy, and smoked snapper in particular is a Kiwi classic. Many old baches have their smokehouse, and most coastal communities have an unofficial 'smokie'. What could better epitomise the Kiwi lifestyle than sitting at the bach or campground, eating a smoked snapper with your fingers, along with friends, family and a cold beer or two?

There are various kinds of smoking. A slow smoke may take hours, including salt-and-sugar curing, then draining and drying taking at least as long again. A true cold-smoke cure can take a couple of days from start to finish.

Most keen fish smokers have their own recipes they have developed over the years. They may involve combinations of temperature, time, types of sawdust and vary greatly in the amount and type of salt, sugar and other ingredients used. Some soak the fish in brine; some sprinkle on the ingredients; others have their own secret additives. Experimenting and developing your own special recipe and technique is all part of the enjoyment, and a smoker who can turn out a top feed is always popular.

THE HOT SMOKE

Hot smoking does not cure fish or extend its shelf-life. It cooks the fish while imparting that lovely smoked flavour to the finished product. It is quick, simple and portable, and provides a delicious meal for very little effort.

The simple and popular way to do a hot smoke is with a portable smoker (usually stainless steel), available from tackle stores. Similar results can be achieved with a lidded barbecue. Because this is a fast-cooking process the fish should not be too thick or the flavour and cooking will not penetrate well. Medium-sized (20–30 mm thick) skin-on fillets, split fish or throats are ideal.

Lay the fish flat on a tray, flesh-side-up, with plenty of newspaper underneath to soak up the overflow fluids. Sprinkle with a layer of salt, then brown sugar. Actual quantities depend on individual taste, but four parts of salt to one of sugar is a good rule of thumb, and half to three-quarters of a cup of salt would be plenty for the fillets off a typical 2 kg snapper. Spread the salt and sugar evenly, making sure you thoroughly cover the thickest parts of the flesh.

Other flavourings can be used to add interest: rum, lemon pepper, Thai fish sauce or sweet chilli sauce are some to try. Now put the tray somewhere cool (away from flies) for a couple of hours. This will allow the salt to draw some of the moisture out of the fish, which will also absorb some of the salt and sugar.

If you are in a hurry you can then put the fish into the hot smoker while it is still wet and covered in residual salt and sugar, and cook it ready to eat within half an hour. Or you can rinse off the excess salt, pat the fish dry and then dry it further in a cool draft of air, again taking care to protect it from the flies. Some people like to use a fan and will dry the fish for several hours, even overnight, until the flesh is dry to the touch. It takes extra time and work to prepare fish this way but the result will be firm and moist and not dripping wet the way fish turns out if it is smoked without the extra drying. However you do prepare your fish, brushing on a little olive oil just before the cooking step will help give it a nice shine. Be sure to spread the oil liberally on both sides of the fins and tail if smoking the fish whole – the result looks better and some people love the fins as a crunchy extra snack.

Most metal hot-smoking boxes come with little meths burners, which can tend to provide an uneven heat, especially in a wind. In still air they work fine but are still a potential fire risk, especially in parched New Zealand summers. A gas barbecue is another good heat source, with the smoke box simply plonked on top.

Setting up is simple. Put a layer of tinfoil on top of the shelf in the smoke box (to make cleaning up afterwards easier), and sprinkle with a handful of sawdust – don't overdo it, or the fish will come out tasting tarry. Some sawdusts are better than others. Natives like kanuka, manuka, and pohutukawa are excellent, as are fruit trees like peach and plum. Do not use pine or macrocarpa, or any treated timber. Sawdust from a chainsaw is also a no-no as it may contaminate the fish with oil.

Put the tray with the fish into the box, put the lid on, then place the whole thing over the barbecue. About 20 minutes is a good cooking time, depending on wind, temperature and fish thickness. Try to keep the heat as low as possible at first so the sawdust doesn't burn off too quickly. Towards the end of the cooking time, poke the fish in the thickest part with a fork to check that it is been cooked right through.

Lastly, slip the fish on to a plate and serve with a crisp salad, some freshly baked bread, and a chilled sauvignon blanc. You won't get a better meal anywhere, I promise you!

snapper

INDEX

BIBLIOGRAPHY

BOOKS

Banks, Joseph, Sir, 1962. *The Endeavour journal of Joseph Banks: 1768-1771* / edited by J.C. Beaglehole. Sydney: Angus & Robertson.

Best, Elsdon, 1929. *Fishing Methods and Devices of the Maori.* Wellington: Dominion Museum.

Blackmore, M.A. [1928?] *The cooking of New Zealand fish and other sea foods.* Auckland: Whitcombe & Tombs.

Brambley, Mavis, 1966. *Sea Cockies of the Manukau.* Wellington: A.H. & A.W. Reed.

Buck, Peter H., Sir [Rangi Hiroa], 1949. *The Coming of the Maori.* New Plymouth: Thomas Avery.

Cook, James. *The journals of Captain James Cook on his voyages of discovery* / edited by J.C. Beaglehole. Cambridge: Published for the Hayklut Society, 1955–74.

Creamer, W.E.H., 1965. *The Tall White Tower.* Unpublished manuscript held at Alexander Turnbull Library, Wellington.

Doogue, Ray, 1967. *Hook, Line and Sinker.* Wellington: A.H. & A.W. Reed.

Doogue, Ray, 1976. *Saltwater Fisherman.* Wellington: A.H. & A.W. Reed.

Evans, Alice, 1981. *Mount Camel Calling.* Auckland: Hodder & Stoughton.

Grey, Romer, 1930. *The fisherman under the Southern Cross: A story of adventure in New Zealand.* New York: Harper Bros.

Grey, Zane, 1926. *Tales of the Angler's Eldorado, New Zealand.* London: Hodder & Stoughton.

Johnson, David, 2004. *Hooked: the story of the NZ fishing industry.* Christchurch: Hazard Press.

Leach, Foss, 2006. *Fishing in Pre-European New Zealand.* Wellington: New Zealand Journal of Archaeology.

Mossman, Sam, 1985. *Saltwater Sportfishing: a new approach for New Zealand anglers.* Auckland: Reed Methuen.

Mossman, Sam, 1992. All You Want To Know About Snapper. Auckland: NZ Fishing News.

Mossman, Sam, 2005. *Sam Mossman's Saltwater Secrets.* Auckland: Reed Books.

Nightingale, Tony, 1992. *White collars and gumboots: a history of the Ministry of Agriculture and Fisheries 1892–1992.* Palmerston North: Dunmore Press.

Orbell, Margaret, and Geoff Moon, 1996.*The Natural World of the Maori.* Auckland: David Bateman. Revised edition.

Poata, Tamati R., 1919. *The Maori as a Fisherman and His Methods.* Papakura: Southern Reprints.

Robson, John, 2000. *Captain Cook's World: maps of the life and voyages of Captain Cook R.N.* Auckland: Random House.

Stevenson, H.K., 1977. *Port and People: Century at the Port of Napier.* Napier: Hawke's Bay Harbour Board.

Sutherland, Temple, 1963. *Maui and Me: a search for a fisherman's Eldorado.* Wellington: A.H. & A.W. Reed.

PAPERS

McKenzie, J. and Holdsworth, J., 1997. Investigation of Snapper Mortality from Recreational Line. Contract Report for the Ministry of Fisheries.

Babcock, R.C., Kelly, S., Shears, N.T., Walker, J.W., Willis, T.J., 1999. Changes in community structure in temperate marine reserves. *Marine Ecology Progress Series* 189: 125-134.

Colman, J.A., 1972. Food of snapper, *Chrysophrys auratus* (Forster), in the Hauraki Gulf, New Zealand. *New Zealand Journal of Marine and Freshwater Research* 6: 221-239.

Crossland, J., 1981. The biology of the New Zealand snapper. *Fisheries Research Division Occasional Publication No. 23.* 14 pp.

Crossland, J., 1982. Movements of tagged snapper in the Hauraki Gulf. *Fisheries Research Division Occasional Publication No. 35.* 15 pp.

Denny, C.M., Willis, T.J. and Babcock, R.C., 2004. Rapid recolonisation of snapper *Pagrus auratus:* Sparidae within an offshore island marine reserve after implementation of no-take status. *Marine Ecology Progress Series* 272: 183-190.

Francis, M.P., 1993. Does water temperature determine year class strength in New Zealand snapper (*Pagrus auratus,* Sparidae)? *Fisheries Oceanography* 2: 65-72.

Francis, M.P. and Pankhurst, N.W., 1988. Juvenile sex inversion in the New Zealand snapper *Chrysophrys auratus* (Bloch and Schneider 1801: Sparidae). *Australian Journal of Marine and Freshwater Research* 29: 565-579.

Hartill, B., Bian, R., Armiger, H., Vaughan, M. and Rush, N., 2007. Recreational marine harvest estimates of snapper, kahawai and kingfish in QMA 1 in 2004–05. New Zealand Fisheries Assessment Report 2007/26. 44 pp.

Parsons, D. and Egli, D., 2005. Fish movement in a temperate marine reserve: new insights through application of acoustic tracking. *Marine Technology Society Journal* 39: 56–63.

Paul, L.J., 1967. An evaluation of tagging experiments on the New Zealand snapper, *Chrysophrys auratus* (Forster), during the period 1952–1963. *New Zealand Journal of Marine and Freshwater Research* 1: 455–463.

Paulin, C.D., 1990. *Pagrus auratus,* a new combination for the species known as "snapper" in Australasian waters (Pisces: Sparidae). *New Zealand Journal of Marine and Freshwater Research* 24: 259–265.

Paulin, C.D., 2007. Perspectives of Maori Fishing History and Techniques. Museum of New Zealand, Te Papa Tongawera.

Richmond, K.S., 1950. On some aspects of the recorded history and the anatomy of *Pagrosomus latus.* Unpublished M. Sc. thesis, University of Auckland.

Scott, S.G. and Pankhurst, N.W., 1992. Interannual variation in the reproductive cycle of the New Zealand snapper *Pagrus auratus* (Bloch and Schneider) (Sparidae). *Journal of Fish Biology* 41: 685–696.

Scott, S.G., Zeldis, J.R. and Pankhurst, N.W., 1993. Evidence of daily spawning in natural populations of the New Zealand snapper *Pagrus auratus* (Sparidae). *Environmental Biology of Fishes* 36: 149–156.

Shears, N.T. and Babcock, R.C., 2002. Marine reserves demonstrate top-down control of community structure on temperate reefs. *Oecologia* 132: 131–142.

Willis, T.J., Millar, R.B., Babcock, R.C., 2003. Protection of exploited fish in temperate regions: High density and biomass of snapper *Pagrus auratus* (Sparidae) in northern New Zealand marine reserves. *Journal of Applied Ecology* 40: 214–227.